'The industrialized designer'

Manchester University Press

To buy or to find out more about the books currently available in this series, please go to:
https://manchesteruniversitypress.co.uk/series/studies-in-design-and-material-culture/

general editors
SALLY-ANNE HUXTABLE
ELIZABETH CURRIE
LIVIA LAZZARO REZENDE
WESSIE LING

'The industrialized designer'

Gender, identity and professionalization in Britain and the United States, 1930–1980

Leah Armstrong

Manchester University Press

Copyright © Leah Armstrong 2024

The right of Leah Armstrong to be identified as the author of this work has been asserted in accordance with the Copyright, Designs and Patents Act 1988.

An electronic version has been made freely available under a Creative Commons (CC BY) licence, thanks to the Austrian Science Fund (FWF) [10.55776/PUB1117], which permits non-commercial use, distribution and reproduction provided the author(s) and Manchester University Press are fully cited and no modifications or adaptations are made. Details of the licence can be viewed at https://creativecommons.org/licenses/by-nc-nd/4.0/

Published by Manchester University Press
Oxford Road, Manchester, M13 9PL
www.manchesteruniversitypress.co.uk

British Library Cataloguing-in-Publication Data
A catalogue record for this book is available from the British Library

ISBN 978 1 5261 4103 3 hardback

First published 2024

The publisher has no responsibility for the persistence or accuracy of URLs for any external or third-party internet websites referred to in this book, and does not guarantee that any content on such websites is, or will remain, accurate or appropriate.

Typeset
by Cheshire Typesetting Ltd, Cuddington, Cheshire

Contents

List of illustrations		vi
Acknowledgements		viii
List of abbreviations		x
	Introduction	1
1	A new profession	27
2	The (general) Consultant Designer	64
3	Women's work	95
4	Professional codes	127
5	Crisis of professionalism	153
6	Social responsibility and the industrial designer	177
	Epilogue	200
	Select bibliography	205
	Index	216

Illustrations

Every reasonable attempt has been made to obtain permission to reproduce copyright images. If any proper acknowledgement has not been made, copyright holders are invited to contact the author via Manchester University Press.

Colour plates

1. Timeline, 'Professionalization of Design in US and Britain, 1930–1980'
2. Gilbert Rohde models a costume from the Saks Fifth Avenue men's shop (1939)
3. Advertisement, 'Olin Cellophane Speeds Package Restyling' (1956)
4. Advertisement, 'Olin Cellophane Sells the Truth' (1952)
5. Advertisement for Du Pont, 'Russel Wright's new plastic design goes nature one better' (c.1955)
6. Advertisement for Smirnoff Vodka, featuring Lucienne and Robin Day (1955)
7. James Ward, cover of *Industrial Design* magazine (1959)
8. Betty Reese with Raymond Loewy, *Boston Herald* (1946)
9. Brochure: 'Industrial Design. New products: source of sales and growth' (c.1965)
10. 'New Space Plan: The Floating Apartment of Mr and Mrs Raymond Loewy' (1955)
11. Gaby Schreiber, hand-painted birthday card (c.1950–1960)
12. Misha Black, two Christmas cards (c.1954–1957)

Table

1   Comparisons of a few leading designers (1943)   68

Figures

1.1 Eric Fraser, illustration of Milner Gray as 'The Designer of the Setting' (1935)   35
1.2 'Mr Designer', 'Britain Can Make It' Exhibition (1946)   39
1.3 James Gardner, Chief Designer of 'Britain Can Make It' (1946)   40
1.4 The Birth of the Eggcup: 'Who designs the eggcup?' (1946)   41
1.5 Workers assembling the 'Cycle of Production', New York World's Fair (1939–1940)   44
1.6 Designers Walter Dorwin Teague, Raymond Loewy, Henry Dreyfuss and Harold Van Doren (1949)   49
2.1 Charts from American Management Association for the Society of Industrial Designers Report (1947)   72
2.2 Diagram showing 'The Designer–Management Relationship in the Corporate Structure' within the Container Corporation of America (1957)   82
3.1 Gaby Schreiber, Consultant Designer, 1919–1991, n.d.   99
3.2 Betty Reese at a Board meeting of Raymond Loewy Associates (c.1950)   108
3.3 '"Pin-up" girl of the month' from *Life with Loewy* (1944)   110
3.4 Dorothy Goslett at the centre of the Design Research Unit (1946)   113
4.1 Advertisement for W. S. Crawford Advertising Ltd, London (1950)   130
4.2 James Boswell, 'Party Conversation' (1952)   131
4.3 Photograph of Peter Muller-Munk, Misha Black and Sigvard Bernadotte, International Council of Societies of Industrial Design (ICSID) General Assembly, Venice (1961)   147
5.1 Cover of Student Handbook, International Design Conference at Aspen (IDCA) (1970)   159
5.2 Ken Garland, *First Things First: A Manifesto* (1963)   171
6.1 'The minimal Design Team', detail from Big Character Poster No. 1: Work Chart for Designers (1969 [1973])   188

Acknowledgements

The research for this book developed from UK AHRC-funded Collaborative Doctoral Award at the University of Brighton ('Designing a Profession: The Structure, Organization and Identity of the Design Profession in Britain, 1930–1980'). In 2017, I received a Smithsonian Baird Society Research Fellowship to pursue a comparative history of the US design profession, which funded the foundational research for this book. In 2020, I was awarded an Austrian FWF Elise Richter senior postdoctoral research fellowship to consolidate and expand this research. I am very grateful to all three of these national research-funding bodies for their generous support of the project, helping it to grow and move in new directions, but particularly for the FWF fellowship, which gave me the time and space I needed to write it.

Thank you to all staff in the following archives where the research was conducted: University of Brighton Design Archives; V&A Archive of Art and Design; RIBA Archives; V&A Museum; Chartered Society of Designers; Royal Society of Arts; Erno Goldfinger Archive, c/o National Trust; J. Paul Getty Trust Institutional Archives; Hagley Museum; National Museum of American History Archives Center and Archives of American Art, Smithsonian Institution; Cooper Hewitt, Smithsonian Design Museum Archives and National Design Library, Rare Books; Hirschl & Adler Galleries and Syracuse University Library Special Collections. Special thanks go to Natasha Bishop, Marge McNinch, Tiffany Miller, Emily Orr and Lesley Whitworth for their exceptional archival research assistance, especially during the lockdown conditions of the global COVID-19 pandemic. Thank you also to the designers and their families who shared private archival documents and personal memories with me in the course of this research: Terence Conran, Ken Garland, Ruth Garland, Anne Hillerman, Horatio Lonsdale-Hands, David Pearson, Alden Spilman, Bridget Wilkins and

Acknowledgements

Susan Wright. I would especially like to thank Anton Bruehl, Jr for kindly providing permission for the use of his father's photograph on the cover and inside the book.

Thank you to all my colleagues in the Design History and Theory department – especially Alison J. Clarke – and to Alexander Damianisch and Franziska Echtinger-Sieghart in the Support Kunst und Forschung department, University of Applied Arts Vienna for their institutional support. Conversations with many people, including colleagues from the University of Brighton, the Glasgow School of Art and the V&A Museum of Art and Design, have shaped the arguments presented in this book. Some of the ideas presented here were also formed in dialogue with others, whose work has inspired me, including: Susan Bennett, Luca Csespley-Knorr, Liz Farrelly, Robert Gordon-Fogelson, Pinar Kaygan, Jessica Kelly, Felice McDowell, Tania Messell, Ali O'Ilhan, Katarina Serulus and Zoë Thomas. I would also like to thank my PhD supervisors Catherine Moriarty and Jonathan Woodham and my examiners, Cheryl Buckley and Sean Nixon.

Further thanks go to the following people who read drafts and offered detailed feedback on this manuscript: Jesse Adams-Stein, Harriet Atkinson, Alison Clarke, Guy Julier, Jennifer Kauffman-Buhler, Tania Messell, Zoë Thomas and Alice Twemlow. I would like to thank Kasia Jeżowska for having the brilliant idea of setting up a Design History Writing group with me and thank all the design historians, including Chiara Barbieri, Julia Błaszczyńska, Yulia Karpova, David Preston, Catriona Quinn, Alyona Sokolnikova and Saurabh Tewari, who provided valuable reflections on this manuscript from its very early stages. It has been a great pleasure to publish as part of the Design and Material Culture Series at Manchester University Press and I want to thank Emma Brennan, Kate Hawkins, Alun Richards and the series editors for their support and expertise at all stages. I would also like to thank the anonymous peer reviewers for their insightful and helpful comments on the manuscript.

Thank you to Janice Spektor for so generously putting me up (and rescuing me) in New York while I was conducting primary research for this book and to all those who have helped with childcare at various stages in its writing. My most heartfelt thanks go to my family, to Freddy and our children Rafael, Melody and Selene, who make us very happy. It is dedicated to them, with all my love.

LA

Abbreviations

ADI	American Designers Institute
AIA	American Institute of Architects
AMA	American Management Association
BSIA	British Society of Industrial Artists
CAI	Council for Art and Industry
CoID	Council of Industrial Design
CSD	Chartered Society of Designers
D&AD	Design and Art Directors Club
DIA	Design and Industries Association
DRU	Design Research Unit
ICOGRADA	International Council of Graphic Design Associations
ICSID	International Council of Societies of Industrial Design
IDCA	International Design Conference in Aspen
ID	*Industrial Design*
IDI	Industrial Design Institute
IDSA	Industrial Designers Society of America
MoMA	Museum of Modern Art
NRIAD	National Register of Industrial Art Designers
RCA	Royal College of Art
RDI	Royal Designer for Industry
RIBA	Royal Institute of British Architects
RLA	Raymond Loewy Associates
RSA	Royal Society for the Promotion of Arts and Commerce
SIA	Society of Industrial Artists
SIAD	Society of Industrial Artists and Designers
SID	Society of Industrial Designers
SCRC	Special Collections Research Center, Syracuse University Libraries
UNIDO	United Nations Industrial Development Organization
V&A	Victoria and Albert Museum

Introduction

> I propose that we name the mid-twentieth century The Age of Disabling Professions, an age when people had 'problems', experts had 'solutions' and scientists measured imponderables such as 'abilities' and 'needs'. This age is now at an end, just as the age of energy splurges has ended. The illusions that made both ages possible are increasingly visible to common sense. But no public choice has yet been made. Social acceptance of the illusion of professional omniscience and omnipotence may result either in compulsory political creeds (with their accompanying versions of a new fascism), or in yet another historical emergence of neo-Promethean but essentially ephemeral follies. Informed choice requires that we examine the specific role of the professions in determining who got what from whom and why, in this age.[1]

Austrian philosopher Ivan Illich's cutting critique of professional expertise, written in 1977, marked a moment of fracture in Western social and political thought, as political activists and critics considered possibilities beyond professionalism and consumer capitalism, drawing attention to the ways in which the two were historically and culturally linked. By the 1970s, industrial design had become one of the most derided professions in contemporary culture. Later, in 1994, Illich would write more specifically about design and its 'pervasive and corrupting influence', as he argued that 'its beginnings can be traced to the rise of modernity and it will almost certainly come to an end with the modern project'.[2] These apocalyptic thoughts on the end of professionalism may seem like a surprising place to begin a history of the industrial design profession. However, the invention, performance and reinvention of the industrial designer, a protagonist of consumer capitalism, presents a vivid study of the short-lived successes and ultimate failures of the professional ideal in twentieth-century history. In Britain and the US, the period from 1930 to 1980 represented a sustained struggle towards professionalization, predicated on planned obsolescence and progress. These essential characteristics, with their imperialist claims to universality, have underpinned our idea of who gets to be called a designer and, perhaps more pertinently, who does not.[3]

This book sets out to interrogate the professionalization of industrial design, making a critical examination of its struggle towards 'omniscience and omnipotence' and its essential failures in fulfilling these goals. Indeed, from one perspective, the history of the professionalization of design is a history of failure and a struggle towards recognition. However, out of this insecurity, the role and identity of the designer was fashioned into something distinctive, flexible and modern; a 'creative profession'.[4] Studying the negotiation, mediation and navigation of professional identity therefore offers a unique perspective on the boundaries, limits and values of professionalization (and its discontents) more broadly. This history of professionalization is particularly pertinent in its telling now, as sociologists, economists, philosophers and designers widely refer to the current moment of 'transition' to 'late' or 'post-'capitalism.[5] The goals at the heart of the Western, industrialized ideal of professionalization – specialization, modernization, technological progress – are increasingly coming under political pressure. As sociologists, cultural theorists and social historians are now making clear, the narratives of work and productivity that drove the twin forces of production and consumption in the twentieth century no longer hold.[6] The automation of work, globalization of markets, societal division and environmental crisis create hostile conditions in which to promote the values of professionalism as universally aspirational ideals. Professional identity is, as such, at a turning point.[7] This, then, would seem like an apt moment at which to look back on the twentieth century and evaluate.

It is precisely the ambiguity of design's professional status that makes it a rich and pertinent case study in the history of professionalization. The designer's status has never been secured according to strict professional lines, despite attempts to do so both by individuals and governmental organizations. It occupies a 'semi-professional' status, alongside other 'new professions' like advertising and marketing. The rhetoric of design, a profession 'which made the public look forward to change',[8] is still fixated on the future. As Guy Julier states, it is a profession 'in a constant state of becoming'.[9] Taking a fresh look at its past to connect with present-day features, this book revisits the history of the design profession in new perspective. Exploring the design profession as a socially constructed practice, the book identifies points of transition, friction and flux that have steered representation and identity in this field since the early twentieth century. Its analysis focuses on the period between 1930 and 1980, starting from the moment British and American designers self-consciously chose to pursue the path to professionalization, to establish visible public status as professionals alongside the architect and engineer. From here, it explores the internal dramas, hopes, aspirations, insecurities and failures of men and women working as designers between 1930 and 1980, a period of immense cultural and social change. It examines the representation of the designer over time on platforms including public exhibitions, magazines and the print press, in the studio, the boardroom and the home. Each chapter of the book captures different moments of transition and agents in the professionalization of industrial design, a profession in a constant state of invention.

Histories of the period 1930–1980 in Britain and the US are frequently punctuated by narratives of transition, from inter-war 'recovery' to post-war 'boom' with the arrival of mass consumption, acceleration of industrialization and, by the 1980s, the advent of globalization and a free-market economy. These transitions were accompanied by major shifts and reorientations in manufacturing and expertise, giving rise to new identities in the so-called 'new professions' of advertising, marketing, public relations, design and management consultancy. Perhaps unsurprisingly, those affected by these cultural, social, technological and economic shifts experienced personal and professional crises, which manifested in a range of issues, from gender to professional self-image – and industrial design was no exception.[10]

This book makes some specific claims on the distinctive value of design as a focus for the study of professionalization in the twentieth century. Hitherto disregarded by sociologists of work and social historians, especially those working within the Durkheimian sociological tradition, design has generally fallen outside the category of what defines a profession.[11] Lacking a universally accepted accreditation scheme in most countries, a commonly accepted code of ethics and conduct and professional representation, the designer's identity has more commonly been associated with the artist, or in later years, 'the creative'.[12] Like others working within this broad cultural category, including advertising and marketing executives, designers pursued careers and identities in the boundaries of 'semi-professional' status.[13] While it has been more commonly historicized in relation to the technological advancement and industrial strength of the machine age, design was also intimately connected to the growth of the so-called 'knowledge' and 'service' economies.[14] This book documents this lesser-known history by identifying the discursive overlaps between management consultancy and design practice in both the US and Britain, through the changing form of design consultancies, the role of consultants within them and the incorporation of management techniques and ideas.[15] It also traces unresolved tensions pertaining to the position of design as a 'new profession', finding that this status meant different things within the distinctive cultural economies of Britain and the US. In both cases, however, the status of the industrial designer was precariously balanced between the boundaries of old and new industry, occupying the newly forged space between production and consumption; as C. Wright Mills would later succinctly put it, 'the man in the middle'.[16]

The 'industrialized designer'

Writing in 1947 in a book to promote the profession to industry, *Good Design is Your Business*, architect and industrial designer Richard Marsh Bennett described the identity of the 'contemporary professional designer':

> This is the designer who has captured the imagination of the public and the confidence of the industrialist ... He is the contemporary name designer, a professional man operating independently, backed by an expert organization combining many talents and contemporary disciplines – the industrialized industrial designer.[17]

Marsh Bennett's use of the term 'industrialized', adopted in the title for this book, successfully captures the dynamics of professionalization in design and its intimate connection to the process of industrialization, technological advancement and masculine identity. As this book will show, agents of professionalization in the US worked to make the identity of the industrial designer synonymous with professionalism, to the exclusion of other design disciplines, including craft and furniture design, thereby rendering non-industrialized practices insufficient and something 'other' to the status of the industrial designer. Professional organizations, including the US Society of Industrial Designers (SID), made this designation on a formal and informal basis, excluding the practice of crafts, textiles, ceramics and furniture from entry to its membership, a position that also came to define the approach of the British Society of Industrial Artists (SIA). The heavily gendered dynamics of the 'industrialized designer' laid bare the hyper-masculine terms on which this identity was formulated, so that it was clear that the professional identity of the industrial designer was a man working in the 'hard-boiled' conditions of the factory or engineering plant.[18] As this book will explore, this masculinity was encoded formally and informally, through the organizations that sought to shape and steer the process of professionalization.

The term 'industrialized, industrial designer' was loaded with professional intent and determination, but it was a term that was still relatively ill-defined. The overdetermined nature of Marsh Bennett's description concealed layers of ambiguity and insecurity. Even the SID struggled to come up with a definition that was satisfactory to its own membership.[19] Meanwhile, another rival organization, the American Design Institute (ADI –later the Industrial Design Institute, or IDI, from 1951), with its origins in the furniture and automobile industries, pursued a definition of the industrial designer that was much more holistic and less 'industrialized'. In Britain, the term 'industrial artist' or 'commercial artist' was more commonly used until after the Second World War. Multiple competing definitions and ideals of what this role entailed – and the practices it represented – were in operation, running alongside one another for most of the twentieth century. For the SIA, later renamed the Society of Industrial Artists and Designers (SIAD), it included and extended to the field of graphic art and illustration, whereas in the US the fields of graphic and industrial design followed parallel paths to professionalization.[20] Nevertheless, as this book explores, by the post-war period, agents of professionalization in Britain seized upon the identity of the industrial design consultant as a model through which to progress the professionalization of the field, settling on a much tighter, exclusive and industrialized self-image. The term 'industrialized industrial designer' also carried geopolitical overtones, as it inferred the advanced status of professions in the *industrialized* over *industrializing* economies. These tensions were exaggerated in later years in the context of cold-war cultural politics, particularly through discussions at the International Council for Societies of Industrial Design (ICSID). Here, the term industrial designer was not translatable across national contexts and was perceived to carry with it the imperialist aims of both the US and Britain, who sought to impose their ideals of professionalization, predicated on consumption and growth on other member states.[21]

Professionalization

Professionalization – the process by which a given field attains professional status in society – has absorbed historians and social scientists since the nineteenth century, producing a rich, substantial and diverse literature to which this book responds and contributes. Design historians have commonly adhered to the structural model of professionalization established by sociologists, by focusing on the establishment of 'professional organizations, educational standards, journals and systems of licensing as instruments through which professions try to define themselves'.[22] This approach builds on the Durkheimian scientific tradition, through which a profession is classified by its structural adherence to organized initiatives and has formed the basis for the historical model of the 'traditional' or 'older' professions of law, architecture, medicine and engineering.[23] These criteria, particularly popular in 1930s sociological studies in Britain, drew a distinction between the 'new' and 'old' professions, an interpretation that was heavily informed by the class system. As Talcott Parsons put it in 1939, professionalism was a mechanism through which normative social power could operate.[24]

Two of the most well-known histories of professionalization in Britain and the US are rooted in socially driven ideas of progress and aspiration. Harold Perkin's 1983 book, *The Rise of Professional Society*, charted the 'inexorable rise' of professionalism as a social ideal in Britain so that by 1970, 'it was accepted in principle by society that ability and expertise were the only respectable justifications for recruitment into positions of authority and responsibility'.[25] Perkin's interpretation of the professional ideal was dependent on the 'interaction of vertical hierarchies based on professional careers, with horizontal hierarchies and class status' and was inherently male-gendered, working to privilege the status of middle-class men in the categorization of professional work.[26] In the US, Samuel Haber's long history of professional culture from 1750 to 1900 argued that 'the American professions transmit, with some modifications, a distinctive sense of authority and honor that has its origins in the class position and occupational prescriptions of eighteenth century English gentlemen'.[27] Haber's book privileges the essential characteristics of professionalism that were marked out by British sociologists and social historians, pointing to the seemingly 'universal' power of professional identity as a gentlemanly ideal 'distinguished by civility, good breeding and manners'.[28]

British historians of professionalism have commonly located the professional ideal to the cultural performance of gentlemanliness. Historian Neil McKendrick defined the operation of the 'professional ideal' as an 'appropriation of gentlemanliness' which guided identities in the 'traditional professions', including architecture, medicine, law and engineering.[29] Feminist social historians Leonore Davidoff and Catherine Hall similarly discussed the equation of gentlemanly masculinity with occupational status in their history of men and women of the English middle class.[30] The association between gentility and professional self-image was perhaps most self-consciously expressed through the identity of the gentleman-architect, an identity which architectural historians have argued was used to disassociate the architect from the artist. As architectural historian Barrington Kaye argued, the British professional classes were suspicious of artists and their claim to professional status. He describes the Victorian notion 'that a profession, even at its best, must be slightly inartistic and that art, even at its best, must be slightly unprofessional'. Victorian society was, he argues, 'divided between men who slept in their top hats and artists who behaved and dressed a class apart'.[31] In *The Image of the Architect*, architectural historian Andrew Saint argues that the architect's professional identity was crystallized in the popular imagination through Ayn Rand's 1943 book *The Fountainhead*, which she dedicated to 'the profession of architecture and its heroes', a statement that captures the spirit of professionalism and its romanticized ethics of individualism and heroic self-determination.[32]

Historians have embraced the sociological distinction between the so-called 'old' and 'new' professions, though more recent work has opened out in the spaces between. Working against this elitist categorization, Zoë Thomas and Heidi Egginton recently sought to use 'gender and marginality as tools to investigate the politics of professionalism', finding that professional boundaries were more porous and permeable than traditionally assumed. Their study illuminated the roles of men and women in the new administrative, technical and commercial professions and argued that 'these gendered histories of success, but also of insecurity, were woven into the very fabric of what had come to be known as professional society'.[33] From this perspective, the 'new' or 'marginal' professions, which include advertising, marketing, design and other commercially oriented practices, have a specific value in highlighting the limits and boundaries of professional discourse. As cultural historian Frank Mort has noted, one of the difficulties confronting the historian of the so-called 'new professions' 'is that the dominant model of expertise has been derived not from the sphere of commerce, but from the genteel and public service professions'.[34] Sociologist Sean Nixon noted the contradictions of this performance in his sociological and historical studies of the British advertising profession, showing that this ideal informed not only how British advertising practitioners saw themselves as 'semi-professionals', but also how they were seen by the government and other professions.[35]

While working within the Durkheimian tradition, sociologist Geoffrey Millerson's 1964 book, *The Qualifying Associations*, offered many insights into professionalization as an informal and formally based process. In particular, Millerson identified the significance of image in the performance of professionalism, which he defined as a 'complex of perceptions, attitudes and beliefs about educational attainment and background, conditions of work, style of life and affiliations and loyalties'.[36] For Millerson, self-image in any given professional field was informed by three principal audiences: the public, the client and other professionals.[37] As the chapters that follow explore, these three social categories took on different levels of agency in relation to the designer's professional identity in the period 1930–1980. Indeed, the decision to privilege the client relationship above all others contributed to one of the ultimate failures of the industrial design profession to establish a real or meaningful relationship with the public, except as a consumer.

Given that design was one of the 'new professions' that arose from new working practices and disciplines formed in relation to consumer culture, it seems important to consider the intersection between cultures of professionalism and consumption, even though these histories have often been written independently.[38] This relationship is crucial in accounting for some of the central failures and contradictions of professionalization in industrial design. Writing in 1947, American industrial designer J. Gordon Lippincott defined the role of the designer as the leader of obsolescence, which he defined as 'the keynote of prosperity'. The designer, he stated, could address the 'major problem of stimulating the urge to buy'.[39] By the 1960s, critiques from both political and environmental perspectives undermined the validity of a relationship between professionalism and commerce and forced designers to turn away from the vulgarity of the market and to 'real world' problems.[40] These new goals, values and aspirations represented a fundamental rejection of everything that had come before. Working alongside anthropologists, sociologists, engineers and psychologists, designers began to reimagine the value of their work in the context of social responsibility. This radical reimagining of the designer's role in society formed the basis for what would optimistically be known as 'social design'. As design historian Lilián Sánchez-Moreno has argued, 'socially responsible design emerged in a European and US context as an attempt to align its discourse with other professions and global concerns'.[41] In this sense, as the final chapters of this book argue, social responsibility was one of the ways in which industrial design organizations attempted to manage the 'crisis' and 'transition' as tools for professional empowerment, with limited success.

Britain and the US frequently professed a certain pride in their 'industrialized' status; a signifier of progress and modernization. Nevertheless, the two nations operated at vastly different scales of manufacturing and production. This was widely acknowledged on both sides of the Atlantic, before and after the Second World War. Industrial design consultants working in the US were convinced that their version of industrial design – and the industrial designer – did not exist anywhere else and that, consequently, it could not be meaningfully compared to any other country, particularly not Britain, where the arts and crafts industries continued to run alongside mass production for most of the twentieth century. Indeed, a great deal of this exceptionalism was built upon the very specific terms of employment that had grown out of the US economy, whereby designers were employed by major manufacturers and corporations on a consultant, freelance basis. By comparison, in Britain, such a relationship between manufacturers and designers was not as clearly defined, and, as the first two chapters of this book will reveal, was more fraught with distrust. Nevertheless, these distinctive characteristics came into dialogue through the emergence of a shared design discourse, which, particularly from the post-war period, played out in the pages of magazines including *Design* and *Industrial Design* (*ID*), in government reports, the debates of learned societies, professional organizations and conferences. The book focuses on these sites of discursive action in the professionalization of industrial design in both places.

The professionalization of industrial design

The professionalization of design was a transnational phenomenon that took on distinctive characteristics within national contexts.[42] Design historians have traditionally adopted the Durkheimian structural approach to professionalization, producing a narrative that begins in Europe, through the formation of a cluster of societies formed to promote the practice and profession of design, including the Svenska Slöjdföreningen (1844) in Sweden and the Finnish Society of Crafts and Design (1874) in Finland. Later, in Britain, the Design and Industries Association (DIA, 1915) and the Double Crown Club (1925) sought to mimic their European counterparts, including the Deutsche Werkbund (1912), to align the aims of art with mass production. By the1930s, a distinctive set of design organizations in Britain and the US signalled a more concerted push to professionalization that was modelled explicitly and self-consciously on the traditional model established through the 'older professions' of law, architecture, medicine and engineering. At this time, designers in Britain, including Milner Gray and Misha Black, set up the Society of Industrial Artists (SIA, 1930), while in the US, John Vassos helped to found the American Design Institute (ADI, 1938) and Henry Dreyfuss, Walter Dorwin Teague and Raymond Loewy established the Society of Industrial Designers (SID, 1944).[43] These societies set out to raise the status of industrial design as a profession in the eyes of the public, government and industry. In both places, professionalization was a social and economic project. As Milner Gray, founder of the SIA, stated, it was driven by the aim of building a self-image for the designer as a gentleman-professional in opposition to the lowly artist. This social aspiration was also implicit in the language of early proponents of professionalization in US industrial design, as Henry Dreyfuss spoke about establishing a 'dignity' for the designer, a new identity that might 'stand almost first' among the professions.[44]

This book began as an effort to test some assumptions within scholarship on both sides of the Atlantic about the relationship between industrial design profession in Britain and the US.[45] As design historians have identified, the exact role transatlantic relations played in the history of design is not always easy to 'pin down', at least at surface level.[46] American design historian Arthur J. Pulos referred to the 'psychological gulf' that existed between Britain and the US in the post-war years.[47] Indeed, at certain points and at key moments, British and US designers sought to diverge rather than interact, particularly on issues of commercialism and ethics. This graphic timeline (see Colour Plate 1) plots the general pattern of professionalization in the two countries, with its rise to intensification in the middle of the twentieth century to declining significance by the 1980s. It visually summarises the initiatives, reports, meetings, publications and exhibitions addressed in this book, which involved the actions of a particular set of agents of professionalization, driven by groups that included government, industry, museums and the media. As this infographic also shows, the course of professionalization was not an even process, but a fluid and ever-changing dynamic. The chapters that follow explore competing ideas of who or what a professional industrial designer should be, according to different agents. Professionalization meant different things at different times to different people, and the effort to establish a common identity across them all formed an enduring struggle for organizations and individuals on both sides of the Atlantic.

Universalism is one of the illusions of professionalism, and this illusion was maintained and performed through a mythologized sense of agency on the part of the individual pioneer. Reporting on the formation of the SIA in Britain in 1934, the *News Chronicle* stated, 'the men and women who have designed the shape of cars and lampshades are gearing up to re-shape their own lives'.[48] In February 1939, US *Vogue* magazine dedicated an entire issue to the industrial designer, 'the men who shape our destinies and our kitchen sinks'.[49] By 1940, historian and designer Arthur Pulos stated, 'individual designers became the glamour boys of the moment – their offices the acme of style ... Magazines also made twentieth-century heroes of the industrial designer and suggested that they may have single-handedly brought the nation out of depression.'[50] The celebrated position designers came to assume as cultural taste leaders placed them, as so influentially stated by sociologist Pierre Bourdieu, in the position of 'cultural intermediaries'.[51] This term was used to describe occupations that typically composed the petite bourgeoisie in the selling of 'symbolic goods and services'.[52] As sociologists David Wang and Ali O'Ilhan have argued, design can be most accurately read as a social practice, where professional identity is formed 'as much through the clothes [designers] wear and home interiors' as through any disciplinary knowledge or professional code of ethics.[53]

Set in a context of 'economic emergency' and 'wartime recovery', design and the identity of the industrial designer was one of the ways in which the state was seeing itself and its future. In Britain, this was markedly shaped by the dominance of the social class system as a determinant of personal and professional identity. As Chapter 1 explores, the idea of design as a 'new profession' formed part of a narrative of national self-determination in both countries. As Gordon Lippincott stated in 1947:

> We Americans are leaders in the sciences and likewise we are becoming leaders in the arts. We are no longer looking to Europe for cultural guidance in the arts, but are boldly striking out with our own ideas. Nowhere is this truer than in industrial design. As a matter of fact, here in America we have the only country in the world where the industrial designer really exists.[54]

This conception of the industrial design profession within a narrative of American exceptionalism forms a recurrent theme. Where US ideas about professionalization made a heavy investment in the future, as Chapter 1 will explore, in Britain, tradition weighed heavy on the minds of design reformers and propagandists. The British insistence on the gentlemanly professional ideal as the dominant narrative lens for the design profession obscures the closely entwined relationship between the designer and the advertising industries, a feature of professionalization there. By contrast, in the US, it is impossible to separate advertising and the history of corporate America from design. Indeed, the corporation looms large as a professionalizing agent throughout this history. As Arthur J. Pulos put it: 'industrial designers define the character of a company to the public on one hand and describe public demands and wishes to the company on the other. They serve as an indispensable link between people and their industries'.[55] Roland Marchand, in his exceptional histories of advertising and the corporation, places the industrial design consultant as a key agent in the 'creation of the corporate soul'. He argues that in the depression decade in which industrial design was forged as a profession, corporations were 'desperately seeking both the trade and goodwill of the common man', an argument taken up in the second chapter of this book.[56]

The history of the professionalization of design in Britain and the US provides another angle through which to test ideas about Americanization, a term used by historian Victoria de Grazia in her thesis of America's 'market empire'.[57] This influential history supported a dominant narrative within design history, whereby the professionalization of British design depended upon the 'fastest possible adoption of the American system'.[58] As design historian Penny Sparke argued in her book, *Consultant Design* (1983), the adoption of the American model of design consultancy and the identity of the design consultant marked the 'arrival' of the profession in Britain. Others have documented the importation of the dynamism and vigour of the American market to Britain mid-century as an important moment of transition from essentially anti-commercial to commercial design work.[59] While researching for this book, however, it became clear that exchange and emulation were resisted and embraced at different moments in US and British design over the course of the twentieth century. In this sense, its findings contribute to recent work in transatlantic studies that move beyond the 'irresistible empire',[60] to present a more nuanced account of cultural hybridization, knowledge transfers, migration and cross-cultural exchange. As Sean Nixon argued in his history of transatlantic relations in the mid-century advertising profession, 'US commercial and cultural influences constituted a resource and stimulus to British advertising practitioners, but one which was reworked and combined with more local cultural resources.'[61] Similarly, historian Jan Logeman has argued that 'transatlantic exchanges in consumer marketing remained multidirectional even as a US style mass consumption appeared to be the dominant global model'.[62] Logeman identifies the role played by elite émigré design consultants, who acted as 'transatlantic mediators', selling the US to the Europeans and vice versa, through 'cross-cultural analysis' techniques.[63] The significance of émigré identity as a feature of professionalization in industrial design, a rich topic explored in several notable research projects and publications, runs throughout the chapters of this book.[64]

Gender

'Hidden histories' of design have tended to focus on the invisible and largely uncredited work of the female designer, but this book contends that hegemonic masculinity and its manifestation in the form of the white, male designer has formed something of an invisible cloak in the design profession. Hiding in plain sight, the relationship between masculinity and industrial design has been under-scrutinized and under-utilized as a tool through which to examine designer identities. This was recently recognized by design theorist Pinar Kaygan, who argued that 'historians have neglected to question the many ways through which technology and business hierarchies have been historically and culturally connected to men and masculinity'. The absence of critical history on the relationship between masculinity and industrial design is a surprising and somewhat troubling realization, considering the extent to which the field of design history has been dedicated to the project of accounting for, documenting and examining the work of male industrial designers. Fiona MacCarthy's well-known history of British design presented a narrative that started with Henry Cole and ended with Terence Conran, posing the question, 'Demi-God or Boffin?' in an ironic, but uncritical way in her 1986 essay on the designer's image.[65] Meanwhile, in the US, the 'pioneering', 'visionary' roles played by Walter Dorwin Teague, Henry Dreyfuss, Norman Bel Geddes and Raymond Loewy have been situated in a narrative of the 'US adventure'.[66]

Writing in 1986, design historian Cheryl Buckley tackled the relationship between design and patriarchy in the early years of the design profession in Britain and called for further work to dismantle the dominance of the omniscient male auteur as the profession's central protagonist.[67] Building on the turn towards masculinity as a method and approach in the humanities in general, this book focuses on the boundary work between masculinity and femininity in the professionalization of design and a strategy through which the limits of professional identity were negotiated. Here, Joseph McBrinn's interpretation of 'hyper-masculinity' as a 'performance of male power through physical spectacle' has been inspiring and instructive.[68] Taking the example of sailors who practised needlework, McBrinn shows how these men used hyper-masculinity to 'negate the feminizing associations of needlework' in Victorian Britain. This approach is illuminating in exploring the formative years of the profession, where representations of industrial design consultants in promotional literature, the media and on the exhibition stage displayed an exaggerated and industrialized image of masculinity. This book argues that industrial designers (and their publicists and promoters) instrumentalized hyper-masculinity to enhance professional credibility, authority and status. This was especially valuable for a profession that had hitherto been represented as 'styling' for industry, an association with feminine overtones. This association with femininity was perceived as a threat by those seeking to professionalize the field; professionalism and femininity had long been considered separate spheres as women were, formally and informally, excluded from professionalized working identities.[69] Technological transitions provoke confrontations between genders and gendered ideals that can manifest in identity crises – and a crisis of masculinity in particular – an occurrence that features in almost every chapter of this book.[70]

Cheryl Buckley's early work on the identity of the ceramics worker in English potteries was also crucial to establishing some of the structural ways in which women were excluded from the profession on a formal and informal basis. Since then, further studies have identified the role of gender in demarcating difference between cultural categories of amateur and professional, especially pronounced in the fields of interiors and textiles.[71] Nevertheless, the figure of the industrial design consultant has been overwhelmingly represented as male. Where women practising in this position have been mentioned, it has been as 'exceptional' success stories, in which their 'diversity and difference' are emphasized.[72] Frequently, women are said to have been 'invisible' or, at best, to have found a way to combine 'femininity with efficiency' to negotiate some form of 'marginal status' within the male-dominated design profession. This uneasy relationship between gender and design history was well articulated in a review article by Judith Attfield in 2003, in which she remarked upon the absence of scholarly research to explore gender beyond 'putting more women on the map'.[73] Instead, Attfield advocated exploring design histories as a negotiation between and beyond gender boundaries. Such an approach, Attfield seemed to suggest, might activate rather than passify the role of women working in the field of design at a range of levels.

Outside design history, breakthrough studies have opened new possibilities through which to reconsider the 'natural' relationship between gender and professional identity. Philippa Haughton, writing on the history of the Women's Advertising Club in London, argued that women took on an active role in the construction of professional identity in advertising in the inter-war years.[74] Haughton's study shows that professional identity in advertising has been 'more fixed in theory than in practice'.[75] Similarly, Zoë Thomas has argued that the home became an empowering site of practice for male and female artists in the early twentieth century.[76] In other more recent work, Thomas has explored the category of marriage in relation to the construction of professional identity for working women artists.[77] Crucially, these studies work to include women within the study of professional identity, rather than focusing on their exclusion from it, and this book takes the same approach. It does this by contributing to what Lesley Whitworth and Elizabeth Darling have termed the 'unknowable woman', looking at the roles played by administrators, managers and publicists within the design profession who were central to its production and performance. This perspective elevates the possibilities for understanding how gender worked in the professionalization of design beyond the limited stock categorization of male and female 'pioneers' and 'geniuses'.[78]

Research conducted for this book uncovered a complex interrelation between genders in the production of professional identity in industrial design. The industrial designer was dependent upon an interaction with the feminized space of consumer culture.[79] As previous studies have identified, the mid-twentieth century saw the rise of the housewife as an increasingly scientific form of expertise.[80] Women's magazines were one of the central sites on which the identity of the industrial designer was first presented to the US public. Male and female industrial designers were frequently pictured in these magazines 'working from home'. Indeed, the home life of the designer – which frequently depended upon a well-balanced marriage where the husband or wife was also an artist or designer – provided a representational device through which to present the aspirational image of this 'new profession'.[81] This book therefore, unexpectedly, has much to contribute to emergent debates about the interrelation of marriage and work, particularly in artistic and creative fields.[82]

Agency in design

'The Industrialized Designer' responds and contributes to the recent shift in design history away from the role played by individual pioneering male auteurs, to represent a more distributed view of professional agency. Many of the designers studied in this book (Russel Wright, Egmont Arens, Walter Dorwin Teague, Henry Dreyfuss, Raymond Loewy, Raymond Spilman, Misha Black, Milner Gray, George Nelson, Freda Diamond, Gaby Schreiber, Terence Conran) will be familiar and well known to design historians, or even to readers outside the field, as 'household names' of mid-century modernist design. As such, their identities are loaded with a certain cultural baggage, which the historian must carefully navigate and account for. It is therefore worth restating here that in giving space to these men and women, the aim is neither to enhance nor discredit their canonical position within design history. Rather, if professionalism and professional identity are taken to be a mythologized performance, the book examines the role played by these actors in performing this myth. This is why the book gives equal weight to the mechanics of mediation – from publicists to professional organizations – through which these actors gained status and significance in modern culture. Furthermore, the question of agency under interrogation in this book is that of professionalization; the object of study is that of the profession, not the designed object.

As the chapters which follow will reveal, the trajectory of professionalization was not even or smooth, as it was an ideal enthusiastically accepted and embraced by only a small minority of those working and practising in the field. It has been accepted as an inevitability of the discipline on both sides of the Atlantic, but empirical research shows that professionalization was a path adopted by a minority group of designers who sought to define, with limited impact in many cases, what it meant to be a designer, distributing power and privilege among a select group.[83] The following chapters challenge this sense of inevitability to show how the profession has been defined as much through the words and actions of those who observed and critiqued it as it was through its most enthusiastic promoters. Moreover, the professionalization of design was in many ways a failed project. As Chapters 5 and 6 reflect, professional discourse in design has been sustained by a sense of crisis, involving tensions between identity, commercialism, gender, design practice and status. Some of the most astute and valuable assessments in accounting for this crisis have come from those identified as 'outsiders' (George Nelson, Victor Papanek and Ken Garland, among others) to mainstream professional discourse. These designers feature as provocative figures who maintained an ambiguous – and autonomous – position in relation to the professional organizations of industrial design in Britain and the US.

While much has been written on the identity and role of the industrial designer as a mediator – or cultural intermediary – in the consumer economy, historians have paid surprisingly little attention to the mechanisms by which designers were themselves represented and mediated to the public. As design critic and historian Alice Twemlow has argued, design historians have tended to take media literature, particularly the 'trade press' publications including *Design* and *Industrial Design*, at face value. As she rightly states, 'magazines encompass the contrary views of, and complex relationships between, publishers, editors, writers, readers, and subjects'.[84] In her history of design criticism in Britain and the US, she draws attention to a dynamic network of individuals and agents involved in a different sort of design work – but one that was absolutely essential to the performance of the profession and to the professional identity of the designer. More recently, Jessica Kelly, in her study of the architectural critic J. M. Richards and her broader research on mediation of architecture, has drawn attention to the 'hidden mechanics' of the architecture profession, which involves the work of critics, magazines, journalists and publicists. The work of these 'mediators' went beyond 'supportive' and administrative roles.[85] Their work was, by necessity and by intent, invisible to the public eye, but crucial to the visibility of the individual 'starchitect' and the promotion of this professional ideal in public life. The politics of display, mediation and visibility form major themes of this study.

Methods

Primary research for this book has drawn primarily on institutional archives from professional organizations in Britain and the US and the archives of individual designers within institutional collections, including the Smithsonian Institution, Syracuse University, the University of Brighton Design Archives and the V&A Archive of Art and Design. As catalogued archives with a certain accessibility and order, these sources present methodological and epistemological questions for the historian, particularly relating to gender and social status. Archives – some of which have been digitized and made publicly available (for instance, in the case of Florence Knoll Bassett) – thereby visibly enhance the status of their subject. This enhanced visibility mirrors and extends the identity of the designer harnessed in public media throughout their professional career. Moreover, as is explored in Chapter 3, the imprint of the publicist or public relations consultant is absent and invisible within this process, even though such people play an integral role in the management of media representation and, in many cases, maintain the press-clippings folder – a source repeatedly drawn upon in this book. Recently, architectural historians have critically examined mediation within archival documentation and illuminated the spaces where the role between archivist and publicist was blurred.[86] This book also makes this argument, building a critical distance from the image of the designer or design organization presented in the archive or biographical study.

Oral history proved to be an invaluable method of exploring critical distance and subjectivity in relation to designer identities.[87] In particular, two oral history collections of interviews with designers, contained within the archives of the Chartered Society of Designers (CSD) and Raymond Spilman's archive at the Special Collections Research Center, Syracuse University Libraries, (SCRC), proved to be invaluable sources, from a number of perspectives. The recordings within the CSD Archive, conducted by designer and past president Robert Wetmore, gave a revealing insight into the gentlemanly cultures of professionalism associated with this Society and its perspective on professionalism.[88] The second set of recorded interviews, in the archive of Raymond Spilman at SCRC, revealed similar issues in relation to masculinity and design. Both sets of interviews, conducted by designers to their peers and for the 'posterity' of the organizations with which they were involved, were highly partial and took the form of a conversation between friends, providing rich material through which to explore questions of self-image and professional identity.[89] Spilman's interviews in particular recorded details on issues such as salary, industry gossip and more polemical debate than had been found in any other archival source. Primary interviews conducted in person with designers and their family members also worked to put past and present issues in the design profession in dialogue.

Geographical limits

In highlighting this particular constellation of professionalization of industrial design (see Colour Plate 1) this book does not seek to reinforce a particular view of the profession, but rather represents a critical examination of its most commonly identified coordinates. Alternative timelines and histories might present the activities of a completely different set of actors, including, for instance, pedagogical, manufacturing or technological advances.[90] Recent work in design history has sought to decentre the dominance of Anglo-American and European narratives. This book contributes to this movement, by providing a critical history of the tools and processes through which cultures of superiority and dominance were constructed. Design, through theory and practice, has always insisted upon its 'universality' in a way that privileges and naturalizes its relationship to actors within the Global North.[91] Through a critical study of two of the most prominent originators of these ideas in Britain and the US, this book contributes to the critical uprooting of this 'natural claim' to universality. Disturbing instances of racial segregation and cultures of white superiority emerged in the course of the research, particularly in the archives of US industrial designers, and a determined attempt has been made to bring this narrative to the surface as part of the fabric of professionalization and modernity.

Similarly, class is an important feature of professionalization that goes deeper than the social distinction afforded to the gentleman-architect, for instance, or in the gentlemanly aspirations of British industrial designers described in Chapter 4. It also arises from structural relations embedded through material conditions of employment, and a study of the scale and scope of manufacturing and industry in both places would bring these issues into sharper relief than has been possible in this cultural study of the professions. Neither is class quintessentially a British 'problem'. As Chapter 5 shows, some of the tensions running high in the 'company versus Consultant Designer' division in mid-twentieth-century US industrial design provide a glimpse into the subtle social hierarchies self-imposed by ambitious and competitive designers, as critic George Nelson cuttingly observed.[92] These hierarchies took on particular pertinence in relation to ideas of freedom and captivity; recurring motifs in the study of professions and creative work.

In offering a critical history of the professionalized identity of the industrialized designer, this book raises the question of professional identity in non-industrialized contexts and outside capitalism.[93] A number of important works have scrutinised the operation of professional identity within the former Soviet states – with fascinating insights into the themes of governance, control, mediation and intermediation that also occur within the analysis in this book.[94] Future studies might reflect on the overlaps and intersections between the values of professionalism and professional identity that were produced in these different social and political contexts. Finally, in examining the histories of design organizations and practitioners in Britain and the US, this study is centred on the activities and identities of industrial designers based predominantly in New York and London, a geographical bias that also skewed the disciplinary and gender dynamics of the field at the time. Professionalism held different values in different geographical locations as connected to disciplinary and industrial concerns. Designers were also aware of cultures of exclusivity and exceptionalism that had been built up around professional practices and identities in these cities. Writing to his Chicago-based peer Dave Chapman, Raymond Spilman wonders if the 'cancer of self-opportunism to the detriment of others is an Eastern trait'.[95] Further research on alternative histories of the design profession in non-urban environments would produce a very different set of questions and issues to those addressed in this study.

Summary of chapters

The analysis that follows takes the form of six chapters, each focusing on particular moments in the construction and mediation of professional identity in design. These adhere to a loose chronology, moving from the 'invention' of the profession in the inter-war years, its mediation through the identity of the design consultant and the controlling authority of the professional organization in shaping professional conduct and behaviour in the profession in the post-war years, through to the crisis of professionalism in the 1960s and the declining significance of professional organizations, both in membership and influence, by the end of the 1970s. However, the book does not commit to a strict timeline, moving backwards and forwards within each chapter to attend to pertinent issues and themes. There is significant overlap between each chapter, as the book zooms in on events, discussions and developments from different angles and perspectives.

The first chapter, 'A new profession', puts pressure on the identity of the 'new', which provided the dominant framing for the introduction of the profession in Britain and the US. The chapter explores the invention of a set of myths, ideals and self-images that guided the profession in its formative years, looking at their circulation and promotion according to different logics of representation and mediation. It argues that while the US professional identity of the industrial designer was defined by visibility, buoyantly celebrated in the pages of newspapers and the trade press, in Britain, the identity of the designer was shaped by its absence, as the industrial artist was said to occupy an 'anonymous status' before the Second World War. While the corporation, individual design consultancies, public relations and media conspired to produce a powerful and colourful image of the individualized designer in the US, in Britain, professionalization was led through governmental agencies and individuals associated with the 'design reform movement', which advocated a more restrained and gentlemanly view of the designer in the image of the 'older professions', including architecture and engineering, resting on a model of teamwork. This tension between new and old status recurs throughout the book.

If the history of a profession starts with the invention of a 'pioneer' through which to sell the professional ideal to the public and business, then the identity of the industrial design consultant represents the apotheosis of this ideal. Chapter 2, 'The (General) Consultant Designer', focuses on the history of the Consultant Designer role, examining its claim to professional status in both Britain and the US. As a central protagonist of the industrial design profession, a title that was celebrated in the US and subsequently in Britain, design historians have previously argued that the 'importation' of this role marked the 'arrival' of professional identity. This chapter reconsiders this assessment, looking more closely at its adaptation and performance, which it finds to be relatively shallow and short-lived in comparison with the movement towards design integration, which represented a constitutive shift in the structure and identity of the industrial design profession in both places. Overall, the chapter argues that the greatest legacy of the Consultant Designer has been the enduring strength of the romanticised ideal of the individual industrial designer.

Chapter 3, 'Women's work', shifts attention away from the male design consultant to look at the production of professional identity on other sites and spaces in industrial design. The chapter begins by examining the identity of the 'woman designer', a term that makes little sense when taken out of the hyper-masculine context of industrial design. Through a focused study of the representation of designers Gaby Schreiber, Freda Diamond and Florence Knoll Bassett, three designers working successfully in the post-war period, it finds that the term was a useful media construction that enabled these women to find a balance between femininity and professionalism, using the former as a tool through which to claim expertise in the realm of consumer goods and interior design. Moving to look more closely at the mechanisms of professionalization, the second section of the chapter addresses the parallel emergence of the publicity profession in the US, as a principal tool through which the visibility and identity of the design consultant was managed and performed. The chapter draws on new research in the archive of Betty Reese, to reveal the gendered dynamics of publicity work and its situation in the office of Raymond Loewy Associates. The final two sections of this chapter examine practices of administration and organization as professional roles in design. Drawing on the freshly transcribed oral history testimony of Dorothy Goslett, Business Manager at the Design Research Unit (DRU), London and a report on 'Student

Behaviour' by Cycill Tomrley, manager of the Record of Designers at the Council of Industrial Design (CoID), the chapter shows how these women formulated their views of professionalism and professional conduct in relation to their impressions of male privilege, which they observed and interacted with at work.

Continuing with the theme of professional behaviour, Chapter 4, 'Professional codes' examines the function of professional societies and organizations as controlling authorities through which ideas about professional identity, its boundaries and limitations, were steered and managed. The first section deals with the establishment of the professional journals as a 'forum for self-definition' through which members could reflect upon the limits and boundaries of professionalism in design. While the chapter finds evidence of this activity in Britain through the pages of the *SIA Journal*, it reflects on the absence of this space in the US, which resulted in a more limited space for self-critique. The chapter then turns to look at the operation of a formalized code of conduct, issued by the respective professional organizations. While each of these adhered to models of professionalism inherited from the 'older professions' of architecture and engineering, the extent to which they were applied in theory or in practice differed considerably in Britain and the US. The professional codes in both cultures were written with the client as their priority audience, while their relationship to the public remained ill-defined and obscure. The chapter focuses on the issue of advertising, which divided opinion almost immediately and became a matter of considerable controversy in both countries. Flagrant diversions from this regulation by some of the profession's most visible protagonists, including Terence Conran and Raymond Loewy, undermined the authority of the professional organization as controlling agents in the profession and by the 1960s, it was clear that the model of professional behaviour inherited from the older professions was incompatible in practice. The final section of this chapter briefly looks at the 'exportation' of the professional code of conduct on the international stage, through the establishment in 1957 of the International Council for Societies of Industrial Design (ICSID), which, it is argued, provided a platform on which British and American designers attempted to project and impose their 'universal' models and codes of behaviour. Subtle acts of resistance to this within the ominously indicated the declining status of these ideals in the context of international politics and social concerns.

Chapter 5, 'Crisis of professionalism', explores the rejection and failure of the professional ideal in design, which originated both from within the profession and outside it. This was predicated through the opening of new spaces for professional dissent, including the International Congress of Design at Aspen (ICDA), which helped to generate and facilitate dissenting discourse that had hitherto been absent from the US industrial design profession. The chapter looks in particular at the 1960 conference, which attempted to bridge a perceived divide between American and British cultures of professionalism, putting the two into dialogue under the title, 'The Corporation and the Designer'. The next sections of the chapter look at internal responses to this growing international critique, as both the SIAD and the newly formed IDSA set out to revise their Code of Conduct in response to increased criticism. As the chapter shows, these organizations proved to be poorly equipped to deal with the scale of this new design culture, as a new generation, favouring cultures of creativity over professionalism, undermined the authority of the professional organization as a controlling authority. Alternative models of behaviour, including Ken Garland's *First Things First* (1963), articulated a cultural change in the professional identity of the designer as it sought to reconcile commercial imperatives with ethical and social concerns.

Chapter 6 turns to the 'reinvention' of the industrial designer in light of 'social responsibility'. Drawing on reports and letters exchanged within the IDSA and SIAD relating to the application for licensing the industrial designer in New York state and the application for a Royal Charter in Britain, the chapter argues that the two organizations were severely inhibited by a poorly established relationship with the public. The chapter examines the public critique of the profession by Victor Papanek, an émigré designer from Austria based in the US, who delivered a damning and dramatic polemic on industrial design, 'a dangerous profession'. The chapter further positions the emergence of 'design for development' paradigms within the context of the cold war and the politicised value of design as a tool of cultural diplomacy and exploitation in industrializing countries. It ends by reflecting on the inability of professional organizations to meaningfully respond to this shift or to sufficiently reinvent their professional identity for this new audience.

The Epilogue reflects on the book's central themes and their relationship to contemporary professional discourse, in which questions of self-definition and identity are central in the profession's ongoing struggle to articulate its value and identity. In this sense, there is a circularity to the book's structure that reflects the circularity of professional design discourse, forever engaged in questions of identity and self-image, forever 'in a state of becoming'.[96]

Notes

1 Ivan Illich, *Disabling Professions* (London: Marion Boyars: 1977), pp. 11–12.
2 Ivan Illich (1994), quoted in Carl Mitcham, 'In Memorandum of Ivan Illich: Critic of Professionalized Design', *Design Issues*, 19:4 (Autumn 2003), 29.
3 Elizabeth Darling identifies the question of 'who gets to be called an architect and who doesn't' as critical to questions of professional status and identity in Jessica Kelly, 'Introduction', Special Issue: 'Behind the Scenes: Anonymity and Hidden Mechanisms in Design and Architecture', *Architecture and Culture*, 6:8 (2018), 7.

4 Industrial designer Ray Spilman identified industrial design as a 'creative profession' in 1961; see Ray Spilman, SCRC. See also Leah Armstrong and Felice McDowell (eds), *Fashioning Professionals: Identity and Representation at Work in the Creative Industries* (London: Bloomsbury Academic, 2018).
5 Jonathan Crary, *24/7: Late Capitalism and the Ends of Sleep* (London: Verso, 2013); Adam Greenfield, *Radical Technologies* (New York: Verso, 2017); Cameron Tonkinwise, 'Design for Transitions – From and To What?', Articles, Critical Futures Symposium, https://digitalcommons.risd.edu/cgi/viewcontent.cgi?article=1004&context=critical_futures_symposium_articles (2015).
6 Andrea Komslosy, *Work: The Last 1,000 Years* (London: Verso, 2018).
7 Andrea Bellini and Lara Maestripieri, 'Professions Within, Between and Beyond: Varieties of Professionalism in a Globalising World', *Cambio*, 8:16 (2018), 5–14. See also Gil Eyal, *The Crisis of Expertise* (New York: Polity Press, 2019).
8 Gordon Lippincott, *Design for Business* (Chicago: Paul Theobald, 1947).
9 '[T]he complexity of assemblages that make up a design culture conspires to propel it so that it is in a constant state of becoming. Design's future orientation gives it a logic of continuous change that makes all objects in a sense unfinished.' Guy Julier, *The Culture of Design* (London: Sage, 2013), p. 248.
10 See Jesse Adams-Stein, 'Masculinity and Material Culture in Technological Transitions from Letterpress to Offset Lithography, 1960s–1980s', *Technology and Culture*, 57:1, 24–53. See also Adams-Stein, *Hot Metal, Material Culture and Tangible Labor* (Manchester: Manchester University Press, 2016).
11 Émile Durkheim, *Professional Ethics and Civic Morals*, trans. C. Brookfield (London: Routledge, 1992 [1957]).
12 David Hesmondhalgh, *The Cultural Industries* (London: Sage, 2002). Richard Florida, *The Rise of the Creative Class* (New York: Basic Books, 2002).
13 Sean Nixon, 'In Pursuit of the Professional Ideal, Advertising and the Construction of Commercial Expertise in Britain, 1953–64', in Peter Jackson et al. (eds), *Commercial Cultures: Economies, Practices, Spaces* (Oxford: Berg, 2000), pp. 53–73. See also Armstrong and McDowell, 'Introduction: Fashioning Professionals: History, Theory and Method', in *Fashioning Professionals*, pp. 1–26.
14 See Guy Julier, *Economies of Design* (London: Sage, 2017).
15 See Chapter 2 of this book.
16 C. Wright Mills, 'The Man in the Middle', *Industrial Design* (November 1958), 70–5.
17 Richard Marsh Bennett, 'The Education of the Industrial Designer', in Richard Marsh Bennett and Walter Dorwin Teague, *Good Design is Your Business: A Guide to Well Designed House-hold Objects Made in U.S.A.* (Buffalo, NY: Buffalo Fine Arts Academy, 1947), CHMRB.
18 Harold Van Doren, *Industrial Design: A Practical Guide* (New York: McGraw Hill, 1940), pp. 80–1.
19 See Chapter 1 of this book.
20 See E. M. Thomson, *The Origins of Graphic Design in America, 1870–1920* (New Haven, CT: Yale University Press, 1997).
21 See also Tania Messell, 'Constructing a "United Nations of Industrial Design": ICSID and the Professionalization of Design on the World Stage, 1957–1980' (PhD thesis, University of Brighton, UK).
22 Gary Beegan and Paul Atkinson, 'Ghosts of the Profession: Amateur, Vernacular and Dilettante Practices and Modern Design', *Journal of Design History* (Winter 2008), 305–13.
23 For this approach, see also Geoffrey Millerson, *The Qualifying Associations: A Study in Professionalization* (London: Routledge 1964). Tony Abbott, *The System of Professions: An Essay on the Division of Expert Labour* (Chicago: University of Chicago Press, 1988).

24 Talcott Parsons, 'The Professions and Social Structure', *Social Forces*, 17 (1939), 457–67.
25 Harold Perkin, *The Rise of Professional Society: England Since 1880* (London: Routledge, 1989), p. 405.
26 Heidi Egginton and Zoë Thomas, *Precarious Professionals: Gender, Identities and Social Change in Modern Britain* (Chicago: University of Chicago Press, 2021), p. 6.
27 Samuel Haber, 'Preface', in *The Quest for Authority and Honor in the American Professions, 1750–1900* (Chicago: University of Chicago Press, 1991), p. ix.
28 Michèle Cohen, 'Manners Make the Man: Politeness, Chivalry and the Construction of Masculinity, 1750–1830', *Journal of British Studies*, 44:2 (April 2005), 313.
29 Neil McKendrick, 'Gentlemen and Players', in Neil McKendrick and R. B. Outhwaite, *Business Life and Public Policy: Essays in Honour of D. C. Coleman* (Cambridge: Cambridge University Press, 1986).
30 Catherine Hall and Leonore Davidoff, *Family Fortunes: Men and Women of the English Middle Class 1780–1850* (London: Routledge, 1987), p. 30.
31 Barrington Kaye, *The Development of the Architectural Profession in Britain* (London: Allen & Unwin, 1960), p. 22.
32 Andrew Saint, *The Image of the Architect* (New Haven, CT: Yale University Press, 1985), p. 9. Ayn Rand, *The Fountainhead* (New York: Bobbs Merrill, 1943).
33 Ibid., p. 2.
34 Frank Mort, 'Introduction: Paths to Mass Consumption: Historical Perspectives', in Peter Jackson et al., *Commercial Cultures: Economies, Practices, Spaces* (Oxford: Berg, 2000), p. 11.
35 Sean Nixon, 'In Pursuit of the Professional Ideal: Advertising and the Construction of Commercial Expertise in Britain 1953–1964', in Jackson et al., *Commercial Cultures*, pp. 55–74.
36 Millerson, *Qualifying Associations*, p. 159.
37 Ibid.
38 See for instance Frank Mort, *Cultures of Consumption: Masculinities and Social Space in Late Twentieth-Century Britain* (London: Routledge, 1996), Erika Rappaport, *Shopping for Pleasure: Women in the Making of London's West End* (Princeton, NJ: Princeton University Press, 1999).
39 Lippincott, *Design for Business*, p. 14.
40 See Victor Papanek, *Design for the Real World: Human Ecology and Social Change* (New York: Pantheon, 1970).
41 Lilián Sánchez Moreno, 'Towards Professional Recognition: Social Responsibility in Design Discourse and Practice from the Late 1960s to Mid-1970s' (PhD thesis, University of Brighton, UK, 2020).
42 See for instance, Chiara Barbieri, *Graphic Design in Italy: Culture and Practice in Milan, 1930–1960* (Manchester: Manchester University Press, 2024). See also Grace Lees-Maffei and Kjetil Fallan, 'Introduction: National Design Histories in an Age of Globalization', in Fallan and Lees-Maffei (eds), *Designing Worlds: National Design Histories in an Age of Globalization* (Oxford: Berghahn, 2016), p. 8. Anna Calvera, 'Local, Regional, National, Global and Feedback: Several Issues to be Faced with Constructing Regional Narratives', *Journal of Design History*, 18:4 (2005), 371–83. Glenn Adamson, Giorgio Riello and Sarah Teasley, 'Introduction: Towards Global Design History', in Adamson, Riello and Teasley (eds), *Global Design History* (London: Routledge, 2011).
43 The American Design Institute was renamed the Industrial Design Institute in 1951 and the Society of Industrial Designers renamed the American Society of Industrial Designers in 1957. The two Societies merged in 1965 to form the IDSA (see timeline in Plate 1).
44 Henry Dreyfuss, *Designing for People* (New York: Simon & Schuster, 1955), p. 222.

45 Leah Armstrong, 'Mimicry, Emulation or Distinction: Mapping Connections between British and US Design Professions, 1930–1960', Smithsonian Baird Society Resident Scholar Fellowship, July–August 2017.
46 Robin Kinross, 'Émigré Graphic Designers in Britain: Around the Second World War and Afterwards', *Journal of Design History*, 3:1 (1990), 35–57.
47 Arthur J. Pulos, *American Design Ethic: A History of Industrial Design* (Cambridge, MA: MIT Press, 1965), p. 224.
48 'Men of Design', *News Chronicle* (1 January 1934), Box 98, CSDA.
49 'Designing Men', US *Vogue* (February 1939), p. 71. See also Leah Armstrong, 'Fashions of the Future: Fashion, Gender and the Professionalization of Industrial Design', *Design Issues*, 37:3 (2021), 5–17.
50 Arthur J. Pulos, *The American Design Adventure* (Cambridge, MA: MIT Press, 1983), p. 9.
51 Pierre Bourdieu, 'The Forms of Capital', in J. Richardson, *Handbook of Theory and Research for the Sociology of Education* (Westport, CT: Greenwood, 1986), pp. 241–58.
52 Pierre Bourdieu, *Distinction: A Social Critique of the Judgement of Taste*, trans. Richard Nice (Cambridge, MA: Harvard University Press, 1984 [1979]), p. 325.
53 David Wang and Ali O'Ilhan, 'Holding Creativity Together: A Sociological Theory of the Design Professions', *Design Issues*, 25:1 (2009), 5–21.
54 Lippincott, *Design for Business*, p. 22.
55 Arthur J. Pulos, *Opportunities in Industrial Design Careers* (n.p.: Vocational Guidance Manuals, 1970), p. 16.
56 Roland Marchand, 'The Designers go to the Fair II: Norman Bel Geddes, the General Motors "Futurama," and the Visit to the Factory Transformed', *Design Issues*, 8:2 (Spring 1993), 17. See also Roland Marchand, *Creating the Corporate Soul: The Rise of Public Relations and Corporate Imagery in American Big Business* (Berkeley: University of California Press, 2001).
57 Victoria de Grazia, *Irresistible Empire: America's Advance through Twentieth-Century Europe* (Cambridge, MA: Belknap Press of Harvard University Press, 2005).
58 Patrick J. Maguire, 'Patriotism, Politics and Production', in Patrick J. Maguire and Jonathan M. Woodham (eds), *Design and Cultural Politics in Post-War Britain: The Britain Can Make It Exhibition of 1946* (London: Leicester University Press, 1997), p. 35.
59 Julier, *Culture of Design*.
60 de Grazia, *Irresistible Empire*.
61 Sean Nixon, *Hard Sell: Advertising, Affluence and Transatlantic Relations, c. 1951–69* (Manchester: Manchester University Press, 2016), p. 4.
62 Jan Logeman, *Engineered to Sell: European Émigrés and the Making of Consumer Capitalism* (Chicago: University of Chicago Press, 2019), p. 3.
63 Ibid., p. 259.
64 See Alison J. Clarke and Elana Shapira (eds), *Émigré Cultures in Design and Architecture* (London: Bloomsbury, 2017). Lesley Whitworth and Sue Breakell, Special Issue, 'Émigré Designers in the University of Brighton Design Archives', *Journal of Design History*, 28:1 (2015).
65 Fiona MacCarthy, 'Demi-Gods or Boffins? The Designer's Image, 1936–1986', in *Eye for Industry: Royal Designers for Industry 1936–86* (London: Lund Humphries, 1986), pp. 17–22.
66 Pulos, *American Design Ethic*; Nicolas P. Maffei, *Norman Bel Geddes* (London: Bloomsbury, 2018).
67 Cheryl Buckley, 'Made in Patriarchy: Toward a Feminine Analysis of Women and Design', *Design Issues*, 3:2 (1986), 3–14.
68 Joseph McBrinn, *Queering the Subversive Stitch: Men and the Culture of Needlework* (London: Bloomsbury, 2021), p. 4.

69 See Hall and Davidoff, *Family Fortunes*.
70 Adams Stein, 'Masculinity and Material Culture'. See also Adams Stein, *Hot Metal*.
71 Grace Lees-Maffei, 'Introduction: Professionalization as a Focus in Interior Design History', *Journal of Design History*, 21:1 (Spring 2008), 1–18. Jill Seddon and Suzette Worden, *Women Designing: Redefining Design in Britain Between the Wars* (Brighton: University of Brighton Press, 1994).
72 For instance, see Liz McQuiston, *Women in Design: A Contemporary View* (New York: Rizzoli, 1988). Elizabeth Darling and Lesley Whitworth critique this representation in *Women and the Making of Built Space in England, 1870–1950* (Aldershot: Ashgate, 2017).
73 Judy Attfield, 'Review Essay: What Does History Have to do With it? Feminism and Design History', *Journal of Design History*, 16:1 (2003), 64.
74 Philippa Haughton, 'Fashioning Professional Identity in the British Advertising Industry: The Women's Advertising Club of London, 1923–1939', in Armstrong and McDowell, *Fashioning Professionals*, pp. 85–102.
75 Ibid., p. 99.
76 Zoë Thomas, 'At Home with the Women's Guild of Arts: Gender and Professional Identity in London Studios, c.1880–1925', *Women's History Review*, 24:6 (2015), pp. 938–64.
77 Zoë Thomas, 'Marriage and Metalwork: Gender and Professional Status in Edith and Nelson Dawson's Arts and Crafts Partnership', in Egginton and Thomas, *Precarious Professionals*, pp. 125–54.
78 Elizabeth Darling and Lesley Whitworth, 'Introduction', in Darling and Whitworth (eds), *Women and the Making of Built Space in England*, pp. 2–3.
79 See Rappaport, *Shopping for Pleasure*.
80 See Lesley Whitworth, 'The Housewives' Committee of the Council of Industrial Design: A Brief Episode of Domestic Reconnoitring', in Darling and Whitworth (eds), *Women and the Making of Built Space in England*, pp. 181–96.
81 See Leah Armstrong, 'Working from Home: Fashioning the Professional Designer in Britain', in Guy Julier et al. (eds), *Design Culture: Objects and Approaches* (London: Bloomsbury, 2019), pp. 131–45.
82 Thomas, 'Marriage and Metalwork'.
83 Claudia Mareis and Nina Paim (eds), *Design Struggles: Intersecting Pedagogies, Histories and Perspectives* (Amsterdam: Valiz, 2021), p. 11.
84 Alice Twemlow, *Sifting the Trash: A History of Design Criticism* (Cambridge, MA: MIT Press, 2018), p. 7.
85 Jessica Kelly, 'Introduction', Special Issue: 'Behind the Scenes: Anonymity and Hidden Mechanisms in Design and Architecture', *Architecture and Culture* 6:8 (2018). See also Kelly, *No More Giants: J. M. Richards, Modernism and the Architectural Review* (Manchester: Manchester University Press, 2022).
86 Beatriz Colomina, *Privacy and Publicity: Architecture as Mass Media* (Cambridge: MIT Press, 1994); Eva Hagberg, *When Eero Met His Match* (Princeton, NJ: Princeton University Press, 2022).
87 See Linda Sandino and Matthew Partington (eds), *Oral History in the Visual Arts* (London: Bloomsbury, 2013).
88 Wetmore appears to have conducted the interviews as part of a broader project to inform the work of gender theorist Liam Hudson, whose work considered the relationship between public schoolboys and creativity among other subjects, providing a fascinating insight into the relationship between social class, masculinity and the professionalization of design in Britain in highly unexpected ways. See Liam Hudson, *The Way Men Think: Intellect, Intimacy and the Erotic Imagination* (New Haven, CT: Yale University Press, 1992).
89 Raymond Spilman, taped interviews, SCRC.

90 Thomson, *Origins of Graphic Design in America*. In her parallel study of the professionalization of Italian graphic design, Chiara Barbieri focuses on education as a major feature of professionalization there. See Barbieri, *Graphic Design in Italy*.
91 Mareis and Paim, *Design Struggles*.
92 George Nelson, '"Captive" Designer vs (?) "Independent" (!) Designer', in *Problems of Design* (New York: Whitney Publications), pp. 28–34.
93 In his talk on the training of the industrial designer, Ray Spilman quotes Yuri Soloviev, who said that the design profession could have better possibilities in a non-competitive society than competitive. 'The training of the industrial designer, 1961–2', Raymond Spilman papers, Talks 1961–2, SCRC.
94 For example, see Triin Jerlei, 'Developing Design Discourse in Soviet Print Media: A Case Study Comparing Estonia and Lithuania, 1959–1968', *Design and Culture*, 12:2 (2020), 203–26. Kasia Jezowska, *Socialist by Design: The State, Industry and Modernity in Cold War Poland*, forthcoming.
95 Raymond Spilman to Dave Chapman (28 June 1960), IDSA 62_9, IDSA Archives, SCRC.
96 Julier, *Culture of Design*, p. 248.

1

A new profession

In 1943, *Life* magazine featured an article entitled 'English Kids', presenting to its American readership the 'home life' of industrial designer Raymond Loewy. The article focused on the Loewy family's temporary care of British advertising art director and designer Ashley Havinden's children, who had come to live in their Long Island home to escape the Blitz. It observed cultural differences between the two families and in the children's tastes, particularly in dress, marvelling at the transformation from their arrival to the country in 'juvenile English fashions' to becoming 'normal self-possessed American teenagers'.[1] Seemingly placed by Loewy's publicist Betty Reese, the feature is a rare example of the relatively undocumented personal relationship between British and US design, through two of its 'leading men'. In focusing on the apparent cultural clash between American and British style, it offers a glimpse into the 'psychological gulf' that characterized the relationship between Britain and the US in the formative years of the professionalization of industrial design in both countries.[2] It demonstrates the extent to which French émigré Loewy's professional self-image was bolstered by his 'transatlantic' contacts and networks,[3] while also showing how interactions between British and US designers frequently served to emphasize difference rather than similarity. The article further serves as a reminder of the significance of war in framing the emergence of the professions in Britain and the US. Borne from economic insecurity, cultural exchange between the two countries was characterized by a mixture of wartime co-operation and national competitiveness.

This chapter investigates the co-construction of an identity for the industrial designer in Britain and the US, focusing on the 'formative years' before the Second World War, which have been subject to a certain kind of mythologization – both by designers at the time and by design historians thereafter. It interrogates the 'invention' of this so-called 'new profession', exploring the representation and identity of the designer through the media, in public exhibitions and professional organizations (governmental and non-governmental). The chapter attempts something of an untangling of these mythologies, through a critical contextualization of the identity and role of the industrial designer as a hyper-masculine hero in the context of inter-war economic recovery and uncertainty, and alongside the advancement of professionalization as a logic of modernization and industrialization. It examines the development of the 'professional ideal' in Britain, finding that it involved a negotiation between new and old traditions of professional culture. By contrast, in the US, design was endlessly represented as a 'new profession', capitalizing on the culture for the new that was embedded in the New Deal ideology of planned obsolescence, a key driver of professionalization there. In the same way that the traditions we think of as 'ancient' have often been 'invented' comparatively recently, historians have learnt to be cautious of the term 'new'.[4] The practice of designing was hardly 'new', with its origins in 'pre-industrial' practices of the arts and crafts. Nevertheless, industrial design continued to be characterized and represented as a 'new profession' in both cultures long after their invention in the inter-war period. This chapter explores the discursive force of the word 'new' in Britain and the US, finding that the term was mobilized in different ways in each place.

Visibility and its relation to professionalization is also a central theme of this chapter. Geoffrey Millerson, a British sociologist of the Durkheimian tradition, developed a theory of professionalization as a 'dynamic process', composed of belief systems, ethics, lifestyles and attitudes. Posing a link between visibility, self-image and the professions, he stated that there are three dynamics to the image of a professional. This refers to the image the professionals hold of themselves (self-image), and how they are seen by the broader public and by other professionals.[5] This chapter argues that self-image and representation are organizing features of the professions and can be read as one way in which professional conduct is made intelligible. The 'formative years' of the profession were significant because, as the chapter explores, a certain image of the industrial designer was taking shape in Britain and the US on the exhibition stage and in the pages of design and mainstream media. The nature of this representation differed considerably in Britain and the US, as it was shaped by different agents and through different media. In the US, the early identity of the industrial designer in the form of a hyper-masculine hero was presented to the public through orchestrated media intervention. Meanwhile in Britain, the identity of the designer as a gentleman-professional was shaped through the agents of design reform, led principally by governmental and voluntary organizations. Nevertheless, both characterizations centred on a display of hyper-masculinity. The chapter is structured in three main sections. For purposes of coherency, the first two sections address British and US

The industrial designer in Britain

Historians have generally understood the inter-war period in Britain as 'transitionary' in the shifting dynamics of work and gender relations. The rising power of print media played an important role here, as new working practices emerged, including marketing, advertising, publicity, design and art direction to service 'mass culture'. Following the extension of the franchise in 1928, collapsing cultures of 'high' and 'low' generated new ideas about power and its distribution across society and sparked a new relationship between producers and consumers. Within this, new professional identities were born; the so-called 'would-be' or 'marginal' professions that were attendant on the growth of new consumer goods in what is now known as the 'cultural' or 'creative' industries. As such, professionalization and the rise of consumer culture in Britain were not parallel concerns, but interrelated events with overlapping and intersecting agendas. The 'industrial artist' – the term most commonly used to identify the 'designer' in Britain in the inter-war period and beyond – bore the brunt of this uneasy transition, alongside other 'precarious professionals' including journalists, advertising practitioners and marketing agents. Social historian Harold Perkin plotted the professional ambitions of these fields within the inexorable rise of professional culture as a working ideal and identity in twentieth-century Britain. However, in Britain, this designation of 'new' status was loaded with derogatory connotations, positioning these professionals as something 'other' to the 'traditional' 'older professions' of law, architecture and engineering.[6] This precarity exposes the significance of social class and status as distinguishing features of professionalization in Britain.

The inter-war years saw the emergence of a plethora of titles and terminologies to refer to the working identities being built in the field of consumer culture, and 'industrial artist' was one of these. As Jonathan Woodham has observed,

> the widely felt uncertainty of the connotations of terms commonly used in the inter-war years, such as commercial art or graphic design, industrial art or industrial design, reflected the inability of designers to establish a clear-cut professional identity or status. Indeed, the use of such terms can lend insights into the changing politics of professional validation.[7]

Industrial art was used to describe a field of work that represented a range of practices, from ceramics to textiles to packing design, and was adopted as a transitory term to identify the work of men and women who practised in this 'semi-professional' space. It was a phrase that no one particularly liked, and yet continued to be used for most of the first half of the twentieth century.[8] The fragility of the term, which attempted to reconcile the seemingly contradictory spaces of art and industry, captures and conveys the precarity and vulnerability that accompanied the emergent profession. Recent scholarship has further problematized terminologies that draw distinction between pre- and post-industrial art practices, arguing that this is a categorization derived from social and cultural construction, rather than any meaningful skill-based analysis.[9] This point has particular pertinence in the case of the professionalization of industrial art in Britain, which had the effect of imposing a divide between London-based industrial artists and textiles and ceramics practitioners in the 'provinces'.

The term 'industrial artist' also engaged with a broader transatlantic dialogue about the place and role of art in industry. Alfred H. Barr's landmark exhibition 'Machine Art' at the Museum of Modern Art (MoMA) in New York attempted to articulate a new relationship between the two, while Herbert Read's *Art and Industry*, published in the same year, opened a reflective space for the collision of these previously disparate concepts to converge. Those working in the field were still viewed with suspicion and snobbishness by cultural commentators, including Wyndham Lewis, who in 1929 described the commercial artist as 'destined to design nothing but smart advertisements for arc-lamps, stomach belts, jumpers, brilliants, the smoke with the ivory tip, fire extinguishers, cosmetics and beautifully curved house-drains'.[10] This cultural bias against the role and status of the commercial or industrial artist was not specific to Britain (it had its advocates elsewhere in Europe and the US too), but remained an obstinate feature of British attitudes to the professionalization of artistic practices.

Design reform

The inter-war period can be characterized as a period of increased 'organized sociability'.[11] Locating the early identity of the designer in Britain involves a careful untangling of promotional design bodies, voluntary and governmental, formed principally between 1915 and 1945 (see Colour Plate 1), which include the Design and Industries Association (DIA), the British Institute for Industrial Art (BIIA), the Council for Art and Industry (CAI), the National Register of Industrial Art Designers (NRIAD), the Royal Society for the Promotion of Arts and Commerce (RSA), the Council of Industrial Design (CoID) and the Society of Industrial Artists (SIA). These societies, organizations and initiatives may be grouped within the design reform movement, a feature of the British government's objectives to boost the country's manufacturing and productive capabilities after the First World War. Design reform and professionalization can be seen as twin, parallel objectives that overlapped at various points in the history of design in the early twentieth century. The nature of their interrelation reveals a great deal about the structure of the British state and attitudes to culture, art and industry that were taking shape under the spirit of a 'new democracy' in the inter-war period.[12]

'Design reformers', as they have been subsequently described, were influenced by activity in Europe, especially Scandinavia and Germany, where organizations including the Deutsche Werkbund had been established in 1907 to improve standards of design in manufacturing and everyday life. In Britain, the DIA was created with similar aims, founded in 1915 by Ambrose Heal, Cecil Brewer and Harry Peach to 'bridge the gap' between art and industry, a frequent refrain of design reform. Although many of the Association's founders, including William Lethaby, were involved in the Arts and Crafts movement, 'the DIA insisted from the outset that it was concerned with industrial products and aimed at infusing a new standard of values into a civilization which was largely based on mass production'.[13] Carrying on in the grand tradition of the 'professional ideal', activity centred around debate and discussion, including lunchtime lectures and meetings, pamphlets and a Yearbook to promote the role of design. DIA lectures usually covered a range of specialist disciplines, from textiles to furniture to pottery, signalling a British preference for specialization that would continue to dominate professional culture for most of the twentieth century. As Herbert Simon, a prominent member, remarked, 'A comforting attraction of the DIA lay in its ability to bring together a group of artists, craftsmen, businessmen and industrial producers and give them an opportunity for discussion and exchange of ideas.'[14] Noel Carrington remembers the Double Crown Club, established in 1924, in similar terms, as a 'club of mostly men', who met to discuss 'the art of fine printing' at a monthly dinner.[15] This gentlemanly culture set the tone for the professionalization of industrial design and the identity of the industrial designer in Britain for most of the twentieth century.[16]

The Society of Industrial Artists

The formation of the Society of Industrial Artists in 1930, by a group of artists and writers in Fleet Street, London, was significant first and foremost as an initiative driven by the aims of industrial artists – specifically those in the fields of graphics and illustration – to advance their status.[17] Although its formative members had been working together for some time as part of the Bassett Gray Group of Artists and Writers, an inaugural meeting at the Café Royal, Soho, London, provided the setting for the Society to announce its professional intent as follows:

> To establish the profession of The Designer on a sounder basis than has been the case hitherto by forming a controlling authority to advance and protect the interests of those who are engaged in the production of design for industry, publishing and advertising.[18]

Representing this as an act of heroic self-determination, in 1934, the British daily newspaper the *News Chronicle* announced the formation of the Society: 'The men and women who have designed the shape of cars and lampshades are gearing up to re-shape their own lives.'[19] While this statement made reference to the one founding female member of the Society – illustrator Lilian Hocknell – the title of the article, 'Men of Design', betrays the dominant view of industrial art as a masculine practice. Hyper-masculine imagery permeated the language used by founding member Milner Gray, when he spoke of the Society's aims to define the image of the designer in opposition to that of the artist: 'Flowing bows, velvet jackets and starvation in a garret is not the paraphernalia of the Industrial Artist. Brains and blueprints are taking the place of these earlier symbols as the artist muscles in on the wider industrial market.'[20] Styling, a term which was never comfortably adopted in Britain, was often smirked at by designers there, because of its commercially driven connotations and the gendered associations with fashion and superficiality. As designers and historians John and Avril Blake commented in 1964, designers in the formative years of the profession were keen to 'dispel the idea that industrial design was the last-minute "tarting up" of a product', a term with misogynistic overtones.[21]

Gray and his fellow founding members were particularly driven towards professional status on the basis of economic, as well as social, improvement. The Bassett Gray Group, founded by brothers Charles and Henry Bassett and Milner Gray, described itself as a 'Group of Artists and Writers', aiming 'to steer a course between the stultifying influence of the commercial art factory on the one hand and the limited opportunities of complete isolation on the other'.[22] The first SIA prospectus laid bare the economic concerns of its members when it noted one of its chief aims was to 'collect debts for members'.[23] Eric Fraser, who worked at Bassett Gray on a freelance basis, regularly designed menus and invitations for the Group's studio parties, called 'bibbing of the patrons', to which clients were invited. The invitation references a 'studio anthem' and parties commenced with the 'taking of the dole', ironically referencing the low status in which its members perceived those working in commercial arts.[24] This gentlemanly sociability formed the basis from which the Society developed its Code of Professional Conduct and guided its principles of professionalism.

Membership of the SIA grew slowly, reaching 410 in 1934, principally composed of London-based practitioners.[25] Speaking at the first meeting of the North Staffordshire Group of the SIA in 1931, Gray made a poignant expression of professional intent when he said, 'I'm tired of today and want to see tomorrow. I need an image not of what I am, but of what I hope to be.'[26] Directed at ceramics workers in the 'Potteries' of England, to whom the promise of professional status organized around collective identity of 'designer' would probably have seemed like a utopian ideal, these words were a poignant expression of the way in which professionalization in Britain was framed as a path to social improvement. Early recruitment outside London, in Coventry and Manchester in 1936, signalled potential to extend and open up debates about professional status and identity in design to a broader constituency, beyond the gentlemen's clubs of London. However, recruitment of female members here was low, reflecting the widely held perception of professional organizations as masculine social spaces.[27] Looking back on the early years in a Presidential Address in 1968, Gray stated:

> I think it would be true to say, and there are few left to gainsay it, that the early fathers of the Society, many of whom worked in the environs of the Law Courts, had become envious of the standing of their friends of the legal fraternity and craved to partake annually of the requisite number of dinners in an Inn of their own, believing that this would enhance their reputations.[28]

Victorian Inns were known to be socializing institutions, instilling cultures of fraternity and sociability in the legal profession in London, rather than providing any legal or technical purpose for the profession.[29] The SIA mimicked in style and identity the various other 'clubs' that had been established to represent and promote design in the format of a gentlemanly monthly luncheon.

Up until the Second World War, the question of how the designer should be represented and seen as a professional was a matter of debate discussed largely within the closed circles of governmental committees and learned societies. In 1936, the RSA made an important contribution to the increased visibility of the individual designer through the establishment of the distinction of Royal Designer for Industry (RDI), which was motivated by the aim to 'enhance the status of designers for industry'.[30] In 1938, a proposal was put before the Council to ask that 'the recipients of the RDI distinction should now form themselves into some kind of Association, with the object of maintaining and advancing the status of the RDIs and of enabling them to act as a corporate body'.[31] Their interests in professionalizing design were formalized in a meeting in 1939, although this remained at the level of meetings, dinners and social events.[32] The Faculty was self-elected, through ballot nominations: 'Names should suggest themselves and not have to be searched for', said the committee at a meeting to discuss nominations in 1950.[33] Members were anxious that the RDI should be representative to the public of the profession of design in all its specialisms.[34] In this way, like the DIA and SIA, the Faculty of RDIs also felt the responsibility to build an image of the profession on a broad basis, specialized into categories of design practice, from textile, to product, graphics to exhibition design.

An 'old' profession

The process of professionalization was, for these would-be professionals, not one of easy assimilation, but rather an awkward form of mimicry, as industrial artists and designers following the 'professional ideal' presented themselves in the image of the gentleman architect. From early on, the SIA was keen to disassociate itself with the 'new profession' of advertising, instead emphasizing their relationship to architecture, an 'older profession'.[35] This was an ambitious aim for an organization dominated by those practising in the fields of graphic art and illustration and it reflected the class-based anxieties of the protagonists involved. Professionalization was viewed as a 'promising method of social reform' in Britain and the older professions of law and architecture served as prestigious models. Design reform advocate Frank Pick optimistically characterized architecture as the 'sister' of design, stating, 'The designer for industry must be placed alongside the architect with a training equivalent in character ... and with a status and authority equivalent too.'[36] The Society's first headquarters were at the Architectural Association (AA), 'through the courtesy of its director Frank Yerbury'.[37] As émigré architect and industrial designer Misha Black put it, designers had to organize themselves as efficiently and competently as 'architects, engineers, barristers, doctors'.[38] The image of the industrial designer in Britain was therefore a composite, drawing on elements of the older professions. This is clearly visible in a caricature of Milner Gray, as the 'Designer of the Setting', by his colleague, illustrator Eric Fraser, Fellow of the SIA (see Figure 1.1).

Caricatured here sporting a top hat, cane and monocle, this playful figure conveys the cultural stereotypes of the gentlemanly professional ideal Gray so self-consciously sought to emulate. In an interview with Gray in 1991, journalist Bridget Wilkins revealed that he often used to arrive at the DRU office in his riding gear, and was referred to by colleagues as 'Hunting Crop Gray'. Wilkins uses the word 'gentleman' four times in this short article.[39]

In tangible terms, the SIA achieved relatively little for the profession before the war, inhibited by lack of government funding or recognition. Indeed, there are some indications that the SIA's ideas about professional identity and status were not well aligned with the government's. In 1936, the CAI set up the NRIAD, an employment agency to liaise between industrial artists and manufacturing companies, without consulting the SIA and with no mention of the Society in its opening report, a snub which was felt bitterly by the SIA's founding members.[40] In its early years, the Society's bias towards graphics and illustration was considered an obstacle to professionalization, a fact illustrated by the Society's stubborn resistance to change its name to include the word Design until 1963. A quantitative investigation of the Society's membership records also reveals a truly holistic approach to membership categories, which were designed to be as flexible and integrative as possible, with designers self-defining in practices that ranged from exhibitions to 'fancy goods'.[41] It was not until after the Second World War, following the establishment of the Society of Industrial Designers (SID) in

A new profession 35

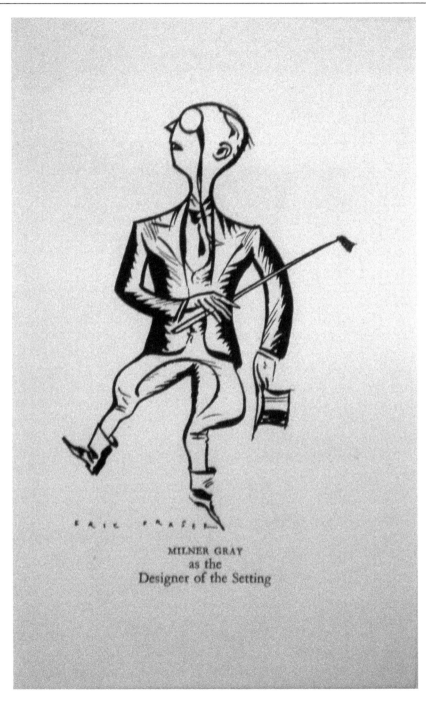

1.1 Eric Fraser, illustration of Milner Gray as 'The Designer of the Setting', Bassett Gray Group of Artists and Writers (1935). Eric Fraser Archive, V&A Archive of Art and Design, AAD/2000/14/72.

the US, that the SIA pursued a more rigid approach to professionalization as an identity for the designer for mass production. Nevertheless, the SIA continued to embrace graphic arts, illustration and design for art direction (2D) alongside industrial design (3D) in its membership well into the 1960s, conforming to the specialized structure of professionalization advocated in the Durkheimian sociological tradition.

Post-war professionalization

Milner Gray announced a new strategy to change the relatively open and informal nature of its membership at a grand post-war dinner held at Claridge's Hotel, London on 11 December 1945.[42] Membership had dropped from 475 to 275 during the war and these 275 were disbanded in the Society's first 'major and very courageous step' to tighten its criteria of membership.[43] The Society built a more rigid membership structure, broken into nine categories (A–H): Industrial Design (A), Design for Craft-Based Industries (B), Textile and Dress Design (C), Display, Furniture and Interior Design (D), Graphic Design (E), Television Film and Theatre (F), Design Direction (G) and Illustration (H). This new structure reflected the specialisms of design, mirrored by other institutions including the RDI. Most significantly, the Society would now exclude crafts-based practices, dividing into two categories of membership, commercial and industrial design, to focus more specifically on those working for mass production. Architect Wells Coates, founder of the Modern Architectural Research Society (MARS) and Fellow of the SIA from 1945, sent a letter of support to Gray the next day, praising the 'arduous task of the reformation of the Society'.[44] The Society's ambitions were clear: to establish the identity of the professional designer as a designer who worked strictly in 'design for quantity production', tilting the Society's emphasis towards industrialization explicitly for the first time.[45] Significantly, this specification worked to overtly exclude those working in arts and crafts and draw a distinct line between amateur and professional according to these highly gendered terms in British design for the first time.

Art Director James de Holden-Stone stated that 'many members' lost their place due to these stricter terms of membership, intimating that this smaller membership base reflected more focused professional intentions.[46] Membership was divided into two groups for the first time: industrial and commercial designers; a distinction that could also be summarized as describing those who worked for mass production in two and three dimensions; the 'flat boys' and the 'round boys', as one designer eloquently put it.[47] The second annual general meeting after the war in 1947 focused on the question of 'whether the Society should seek to make itself representative of technically competent practising commercial and industrial artists or alternatively represent only those who have achieved a relatively high aesthetic in addition to a technically competent standard'.[48] The debate concluded that the Society should 'maintain its rigid membership structure' in favour of further advancing the status of the profession. Designer and design reform propagandist John Gloag stated that the recent formation of the Society of Industrial Designers (SIA) in the US, a society modelled on strict entrance requirements guarded by its founding members, was a 'significant factor' in this tightening of the SIA membership structure, a statement which is difficult to verify. The formation of government committees and the publication of major governmental reports that were directed at the tightening of professionalism as a strategy for improving standards in manufacture and production, along with the formation of the government-sponsored CoID, were likely to have been more direct influences. Here again, the context of post-war economic recovery and reconstruction following the Second World War formed a crucial backdrop for this new push towards professionalization in design, with an emphasis on mass production to stimulate consumer culture and spending.

'Mr Designer'

While the immediate post-war period was marked by increased activity in the promotion and representation of design as a professional activity in Britain, there is little evidence to suggest that this made any impact on the general public. With this in mind, several government reports listed exhibition strategy as an appropriate method of illuminating the designer's role, and this was first realized through 'Britain Can Make It', staged at the Victoria and Albert Museum, London in 1946. Co-ordinated by the newly established CoID, 'Britain Can Make It' was one of the most visited exhibitions of design ever staged in Britain, attracting record visiting figures.[49] The exhibition can be viewed in the context of design reform as part of the aim to promote the value of Good Design in Britain, with the tagline 'Good Design and Good Business'. Over five thousand items were on display, encompassing furniture, tableware, domestic appliances, household equipment, carpets, wallpapers, clothing and toy design. As design historian Harriet Atkinson states, the exhibition was 'not simply focused on selling goods, but on demonstrating wider ideas about Britain's productive capabilities, with the aim of explaining the technical aspects of industrial design to the public'.[50] This was further enhanced by the recent establishment of the Council's photographic library, which instrumentalized the Council's representative capacity through the medium of photography.[51] While the exhibition put on display the work of the most prominent industrial designers working at this time, the CoID also appointed three 'fledgling' designers to design the exhibition: James Gardner, Basil Spence and Misha Black.[52] All three were founding members of the SIA, while Spence and Black were also trained architects and members of the RIBA. In the context of relative anonymity, the exhibition presented these men with a stage on which to launch an image of their profession to the public for the first time.

Responding to this brief, both Gardner and Black put the individual designer centre stage in the exhibition, elevating the status of the designer to one of authoritative expertise, using theatricality and drama to maximize impact. Gardner's 'Mr Designer' narrated the entire exhibition and directed the audience around the site, producing an unforgettably strange assemblage, composed of an 'all-seeing', enlarged eye with jagged eyelashes and commanding, outstretched hand which spoke to the audience through a loudspeaker, guiding them through the exhibition, prompting their expectations and responses (see Figure 1.2). Putting the designer in this central narrative position, Gardner made an ambitious, confident assertion of the designer's role in Britain's post-war economic recovery. As design historian Lesley Whitworth put it, the figure 'boldly holds our gaze, born out of wartime ingenuity, and blazoning the way into the future'.[53] Whitworth also comments upon the hyper-masculinity of the image:

> His shiny angularity and all-seeing eye humanizes the enterprise that came to fruition in the gloomy basement spaces of the Victoria and Albert Museum, and gives this austerity-era offering a heroic aspect that its later and more popularly acclaimed successor, the 1951 Festival of Britain, arguably lacked.[54]

'Mr Designer' was also a manifestation of the didacticism of post-war design reform, characterized by initiatives such as the 'Design Quiz', devised by the CoID and also featuring in the 'Britain Can Make It' exhibition. The image of the eye permeated the CoID's promotional activities and initiatives, building on associations with insight, expertise and vision, motifs of modernity that had been promoted by graphic designers of the Bauhaus and beyond, including György Kepes.[55] It is striking to note the visual coherency between 'Mr Designer' and the photographic portrait of James Gardner, held by the CoID within its photographic library, staring at the viewer in what could be read as a bold and determined expression of professional intent. The pipe, a signal towards the gentlemanly eccentricity of the designer's identity, was a common feature of photographic portraits of designers in this period (see Figure 1.3; also Figure 4.2).

In addition to 'Mr Designer', 'Britain Can Make It' opened up a space to introduce the industrial designer to the public for the first time. Section 16 of the exhibition, 'What Industrial Design Means', set out to explain to the public, in visual form, 'the issue of what industrial design encompasses'; a 'rather large question', as designer Norbert Dutton admitted.[56] Black, an architect and designer known for his inventive use of narrative techniques in exhibition display, used a storyboard to direct the visitor on the process of designing an eggcup. The script, written by Black, was explicitly gendered: 'Here is the Man. He decides what the eggcup shall look like. He is the Industrial Designer. He works with

A new profession 39

1.2 Mr Designer, 'Britain Can Make It' Exhibition (1946). Designer: James Gardner. Photographer: unknown. Design Council Archive, University of Brighton Design Archives, BD-1837-DES-DCA-30-1-13-13-7.

the Engineers, the Factory Management – and is influenced by what you want.'[57] Black invested a great deal of himself in his dramatization of the industrial designer's role. Remembering her first day as Design Manager in the DRU office, Goslett describes encountering Misha Black

1.3 James Gardner, Chief Designer of 'Britain Can Make It' (1946). Photographer: unknown. Designer Portraits, Design Council Archive, University of Brighton Design Archives, GB-1837-DES-DCA-30-1-POR-G-4-5

in a room 'full of egg cups, of all sorts and size'.[58] Black continued to use the eggcup analogy to define his role throughout his career, stating on British television in 1960: 'I myself am an industrial designer and jobs which I have done range from the design of small plastic egg-cups to the bodies of huge steel locomotives', emphasizing the range of his work in a similar manner to his US counterparts.[59] While Whitworth and others have suggested that the selection of this seemingly mundane everyday household object as the central representational device through which to explain the industrial designer's role was essentially arbitrary, it also seems possible that it was selected to connect the housewife – the 'ideal middle-class consumer' in American parlance – with industrial design.[60] Indeed, in another educational booklet Black wrote to accompany the touring exhibition, he compared the designer's use of decoration to the woman's use of make-up.[61]

In its muscularity and bold posturing, with commanding, outstretched hands, Black's 'Industrial Designer' mirrors the 'industrialized' hyper-masculinity of 'Mr Designer'.

A new profession 41

1.4 The Birth of the Eggcup: 'Who designs the eggcup?', 'What Industrial Design Means' section of the 'Britain Can Make It' Exhibition (1946). Designer: Misha Black. Photographer: unknown. University of Brighton Design Archives, GB-1837-DES-DCA-30-1-13-30-34.

British design historian Jonathan Woodham has been dismissive of the overall impact of the 'What Industrial Design Means' section of the 'Britain Can Make It' exhibition, citing evidence from CoID survey questionnaires which show that the public rarely mentioned it in their responses.[62] Nevertheless, as architectural historian Elizabeth Darling has noted, the exhibition was significant for putting the 'fledgling profession' of industrial design on a platform for the first time.[63] Furthermore, while the exhibition as a whole might not have held much significance for the general public, anecdotal evidence in the form of oral histories and biographical accounts suggests that the 'What Industrial Design Means' section in particular did have a significant impact on a generation of aspiring professional designers who saw the exhibition, including British designer David Carter, who specifically remembered seeing the exhibition in his 'bell-bottom trousers ... standing in front of Misha Black's stand with the man with the egg cup' and saying, 'That's what I want to do ... That's the first time I have seen it described.'[64] Revealingly, contemporary critics were concerned that the 'Birth of an Egg Cup' section placed too much emphasis on the individual designer, misrepresenting the practice of design as a one-man show. In a report to the CoID in 1947, Noel Carrington said, 'in too many places it reads like a boost for the profession of industrial designer', revealing characteristic wariness on the part of British design propagandists who wanted *design* and not the individual *designer* to take centre stage.[65] In this sense, it can be viewed something of a coup for Black, Gardner, the DRU and the agency of the designer.

A new profession in the US

In the US, the economic 'New Deal', a series of relief reforms between 1934 and 1939 which aimed to revitalize the economy principally through consumer spending under President Franklin D. Roosevelt, provided the stimulus for the growth of new 'professions' and practices, including industrial design.[66] Ideas about the 'consumer problem', as it was articulated by sociologists, consumer activists and liberal economists in the 1930s, were linked to new conceptions of democracy and citizenship. As historian Meg Jacobs put it, the production of 'well-informed, independent and financially secure consumers' was central to the prevention of 'economic ruin and civic decay', a fear that was stoked in the context of the Great Depression and the threat of fascism in Europe. The growth of the 'middle-class ideal' through the 'stylization' and extension of the middle-class consumer market drove this process.[67] The industrial designer emerged in the context of this economic programme as a protagonist of the corporation, selling consumption and planned obsolescence to consumers, often defined as 'middle-class women, thought to be irrational and endlessly suggestible'.[68]

Whereas in Britain, promotional efforts had focused on the value of 'Good Design' to improve standards in manufacture and production, in the US, the 'newly minted' industrial design profession was predicated on planned obsolescence, a condition of constant reinvention, as an aid and stimulus to consumer spending.[69] Within this context, a narrative was assembled between public relations experts and corporations that put the industrial designer at the centre of a heroic economic rescue. The New York World's Fair played a major role in launching the identity of the industrial designer to the public. The title for the Fair, 'Building the World of Tomorrow', was an apt one for a profession in a state of self-invention, and much of the literature used to promote the Fair focused on the role played by industrial designers. Context had been set at the Chicago World's Fair in 1933, where 'the guiding hand of the industrial designer had transformed the older exhibits of a static conventional character into compelling displays that drew increased thousands on whom they left indelible impressions'.[70] Promotional material to accompany the New York World's Fair put designers at its forefront, using them as protagonists to sell the message of progress, technological innovation and 'the future' to the American public.

This sculpted display of automotive parts and shiny cars, stacked high upon one another by the exhibition designers (Figure 1.5), captures the spirit of planned obsolescence and its centrality to the new profession. The image holds striking contrasts with that shown in Figure 5.1, discussed in Chapter 5, which visualizes the dramatic collapse of the profession, as it came under popular critique by the 1960s.

The Fair's publicity office, directed by publicist Louise Bonney-Leicester, positioned industrial design consultants as protagonists through which to promote the Fair itself, and their masculinity was central to this representation. The February 1939 issue of US *Vogue* was dedicated exclusively to the Fair and the industrial design consultant received special attention in a feature 'Designing Men'. The language of the piece shaped a hyper-masculine narrative in what would later be dubbed the 'heroic age of capitalism'.[71] According to this narrative, the professionalization of industrial design became part of the 'American design adventure'.[72] Elements of this narrative were already in place before the Fair. In a widely cited editorial in *Fortune* magazine (February 1934), anonymously penned by industrial designer and critic George Nelson, industrial design consultants were presented in heroic terms as leaders of a 'serious, new profession'.[73] As design consultant Brooks Stevens put it in his annual address to the American Institute of Electrical Engineers in 1956, the industrial designer's entrance to professional life was intrinsically linked to the economic depression:

> At this point, the industrial designer and the sleeping profession emerged and 'operation bootstrap' began. The recognized pioneers set forth, as individuals or with small staffs, to revitalize selling by heavy emphasis on eye-appeal and buy appeal ... we were on the brink of new materials, new tooling approaches and new horizons of sales volume that would provide dollars for experiment, exploitation and better products for the woman's world.[74]

'New-ness' had a powerful discursive force for the professionalization of industrial design. Whereas designers and agents of design reform in Britain had focused on the challenge of building the new profession of design in the image of the older profession of architecture, in the US, media editorial and career guidance literature seized on the cult of the 'new' that had gripped inter-war public discourse. This reference to the 'new' served

'The Industrialized designer'

1.5 Workers assembling 'Cycle of Production', Ford Production Cycle exhibit, New York World's Fair (1939–1940). Manuscripts and Archives, MssCol 2233, New York City Public Library.

more than to designate its 'recent' invention. Industrial design was also a new *type* of profession, designed to sell 'the new'. Indeed, industrial design was represented as a 'new profession' well beyond the Second World War, as later chapters of this book will explore.

Writing in 1947, Gordon Lippincott said, 'It is probably true that about two new cars out of five are bought each year because of style obsolescence. If appearance improvement each year can sell more American merchandise, then the industrial designer is playing a key position in maintaining increased employment ... The industrial designer has made the public look forward to change.'[75] Standing at the intersection of colliding agendas of professionalization and consumer culture, the 'new profession' of industrial design represented what cultural historian Jan Logeman has described as the 'new typology' of 'consumer expert' in the US, alongside the marketing and audience research consultant, 'consumer engineer', advertising, publicity and public relations.[76] As professions 'requiring new forms of scientific and aesthetic knowledge', Logeman argues that these experts became 'a driving force of mid-century transformations in consumer capitalism'.[77] US industrial designers also posed a direct relationship between their own profession and the 'new professions', including advertising. Speaking in 1953 at the American Association of Advertising Agencies, Henry Dreyfuss, President of the SID at this time, described his 'fledgling profession' as a 'natural ally to advertising', with a 'unique mutuality of interest', which included the fact that they were 'both creatures in mass production'; 'both measure[d] the value of our services in sales'; both professions required a keen sense of timing, having to know 'what to sell and when', needed a 'sixth sense about public taste' and should be 'interested in client's profits'.[78]

'Designing men'

The discursive effects of gender in the professionalization of industrial design were played out in a special issue of US *Vogue* in February 1939, dedicated to the New York World's Fair and entitled 'Fashions of the Future', in which nine industrial design consultants were invited to contribute their vision of the 'woman of tomorrow'.[79] This was a significant promotional opportunity for the industrial designer to present their new profession to a predominantly female readership, thereby accessing the 'ideal consumer' of the New Deal economic agenda.[80] Editor Edna Woolman-Chase introduced the feature:

> The men who shape our destinies and our kitchen sinks, streamline our telephones and our sky-scrapers, men who brought surrealism to Department Stores and be-Thyloned Perisphere to Long Island. They know about the problems, the dreams and the realities that the future has in store for us. They are trained to think ahead: they know tomorrow like they know their own streamlined pockets. We will offer them the hospitality of our pages and let them have some fun with the clothes of tomorrow.[81]

This idealized vision of the industrial designer bore very little relation to reality. For one, the reference to scientific 'training' could be challenged on the basis that industrial design education was at an emergent stage in the US.[82] As George Nelson pointed out in his portrait of the profession in

Fortune (1934), most of its leading protagonists had no educational training at all, having moved into industrial design work through experience and contacts in related fields including illustration, art direction and theatre design.[83] The fantastical visions submitted by each designer functioned in two main ways: first, by promoting the corporation each designer was working with at the Fair and secondly, by reimagining fashion as a rationalized, industrial discipline, where 'feminine' values of style, taste and glamour are rejected in favour of function, order and innovation. Industrial designer Donald Deskey, for instance, wrote of how women would wear a 'system of clothes units'.[84] The value of fashion skill was diminished by the designers, who instead sought to present industrial design as a discipline of the future. Materials and fabrics would be 'blown' or 'rolled out like cellophane'; stitching would be replaced with 'a cementing or moulding process'.[85] This hyper-masculine rejection of fashion was in keeping with modernist design principles, which worked to discredit the values of fashion, underpinned by acclaimed philosophical texts such as Horatio Greenough's 'Form and Function', widely read across architecture and design.[86]

In her letter to the readers, Woolman-Chase explained that 'some of them admitted that they have never designed anything more frivolous than a locomotive', while in perhaps the most overt dismissal of femininity, one industrial designer, Gilbert Rohde, 'said he would much rather design a costume for the Man of the Future'. Rohde's vision (shown in Colour Plate 2) presents a striking image that appeared to borrow ideas from the men's rational dress movement of the period, which he directly references as the 'Great Revolt'.[87] In a combination of dramatic, theatrical posturing and steely angularity that characterized the work of its photographer Anton Bruehl, Rohde's stance in the image is commanding and authoritative: a truly 'industrialized' man of the future.[88] The detailed description that accompanies the image shows the serious consideration Rohde gave to the values of masculinity, while simultaneously revealing the insecurities that underpinned his bold proclamations. 'Women's clothes are pretty good as they are,' he said, 'but men's need radical revision.'[89] Rohde, who himself wore a beard, was particularly exercised on the question of facial hair, which he seemed anxious to defend for the man of the future, an issue that recurs with surprising regularity in the chapters of this book.

The text accompanying the image read:

> Man of the next century will revolt against shaving and wear a beautiful beard, says the designer of metal furniture, lamps, clocks, pianos, boilers. His hat will be an antenna snatching radio out of the ether. His socks – disposable. His suit minus tie, collar, buttons. His belt will hold all his pockets ever did.[90]

This image of the 'man of the next century' shares some similarities with 'Mr Designer' presented at 'Britain Can Make It', in the steely industrialized, masculine aesthetic and bold, assertive posturing. Together, these hyper-masculine images, with their exaggerated authority of 'omniscience', can be read as a potent expression of professional intent. Nevertheless, they also hint towards underlying insecurities in the designer's position in society in relation to both gender and design practice and highlight the frequent association of ideas of masculinity within professional design discourse, particularly in relation to questions of identity and status.

Professional organizations in the US

In a report on industrial design standards in England, Edgar Kaufmann, Director of Industrial Design at MoMA, New York, reported upon English design activities, which have 'caught the interest of the American design world'. Kaufmann explained how, 'from an American viewpoint, government involvement is mirabile dictu', but explained that 'England is small and close knit; government and trade mesh habitually.'[91] While Kaufmann was accurately referring to the absence of governmental support through directly funded organizations or initiatives, it should also be said that the New Deal programme operated through the corporation and in the media, steering the promotion of the industrial designer at this time. Designers themselves also self-organized through 'consciousness-raising' organizations, activities and groups (as illustrated on the timeline in Colour Plate 1), especially in urban centres including Chicago, Detroit and New York. In a similar vein to the DIA or Double Crown Club, the American Union of Decorative Artists and Craftsmen, founded in 1928, has been widely credited with raising the status of the designer through socially centred events, including exhibitions, talks and even dances, to which European designers were invited. In an interview with Arthur J. Pulos, Raymond Spilman was particularly enthusiastic about these events, which he said inspired US designers to form networks and make connections that helped to improve standards of their work and build a sense of common purpose and collective identity.[92]

In 1936, the American Designers Institute (ADI) was founded by a group of designers working mostly in the field of furniture design in Chicago, expanding to represent the field of automotive design in Detroit. Until 1944, when a New York chapter was established, the Society had low representation there and was not seen to directly address the specific concerns or aims of Consultant Designers. Members were encouraged to list their specialisms; Alfons Bach, past President, listed a total of 21 specialisms under his name in the membership roster for 1945; Freda Diamond listed 14.[93] In a letter to Norman Strouse, founder John Vassos described the IDI as 'the oldest and largest industrial design organization in the country and [it] represents the key designers and design executives of America's major industries involved in the design of products'. Vassos was especially proud of the density of membership in Detroit, stating, 'we have in our membership and leadership every major design executive in the automotive industry'.[94] In addition, the majority of the ADI's membership was composed of designers who worked from 'within' companies, organizations and corporations as in-house staff designers alongside a smaller proportion of independent design consultants. Female members were a minority, but present in a range of disciplines and levels. Indeed, women are particularly prominent in the ADI newsletters during wartime, taking on issues including education, competition and licensing of the profession, as reported in the April 1945 edition.[95] The ADI would go on to establish chapters in Boston, Central New York, Chicago, Detroit, Florida, Los Angeles, Ohio Valley, San Francisco and South New England. It also put great emphasis on its interaction with student members and was active in the invention of an innovative and internationally praised educational curriculum for design, through William Katavolos at Parsons New School for Design, New York. Members of the ADI were especially proud of a review by British design critic John Blake published in *Design* magazine in 1963, for his praise of this report, a 'most comprehensive and well-argued analysis of the nature of industrial design'.[96] In this sense, the organization made a significant and often undervalued contribution to the professionalization of design in the US.

The SID was founded in New York by four industrial design consultants: Walter Dorwin Teague, Raymond Loewy, Henry Dreyfuss and Harold Van Doren. As the composition of the image shown in Figure 1.6 conveys, these four designers sought to self-consciously build an image of expertise and influence in the media. Drawing on connections within US corporations and a budget for media publicity, these formative members of the SID secured a visible and prestigious position within the historiography of the industrial design profession in the US. However, the history of the Society's formation arose from a narrower set of interests and a smaller membership than the ADI. Early attempts at co-operation on issues of mutual concern, namely licensing, faltered.[97] While the SID publicly denied its efforts to 'poach' members from the ADI for its own membership, privately, the archive reveals many such instances.[98] This bitter division between exclusivity (SID) and inclusivity (ADI) shaped the trajectory of professionalization in US industrial design, characterizing the division between consultant and corporate designer, explored further in the next chapter. Neither organization received any governmental recognition or support until well into the 1960s.

Writing in 1962, Arthur BecVar summarized two types of professional organization: 'large, which admits members on the basis of certain minimum qualifications and encourages them to agree to some self-imposed, broad standards of ethics … the other is a small, elite, high prestige group, which is composed of the most illustrious members in a profession'.[99] BecVar was referring to the distinction between the ADI

1.6 Designers Walter Dorwin Teague, Raymond Loewy, Henry Dreyfuss and Harold Van Doren, *Time* magazine (31 October 1949). Photographer: Roy Stevens.

and SID. In his oral history of the profession, designer Raymond Spilman acknowledged a degree of exclusivity to the SID's early aims, stating that 'Walter [Dorwin Teague] thought that the Society should be small and exclusive.'[100] Spilman recounts that his recommendations to invite furniture designer Charles Eames and designer, architect and critic George Nelson to join were not considered acceptable to the rest of the group on the basis that they specialized in one field of design activity.[101] The elitist, inward-looking culture of the Society is clearly expressed in the language of its early publications and meetings. New York formed the focal point for the SID, because as Van Doren explained at a Board of Directors Meeting in 1948, 'it so happens that there are more people practicing Industrial Design in New York City and its environs than anywhere in the world'.[102]

There was also a more practical reason for the SID's focus on New York, as the formation of the SID grew from a very specific concern over taxation. New York Consultant Designers Dorwin Teague, Dreyfuss and Loewy wished to receive the same favourable tax status as self-employed professionals, including architects and doctors, in New York state. The three made a joint agreement not to pay the unincorporated business tax liable to self-employed non-professionals working within the state of New York, and encouraged others employed in a similar status not to do so.[103]

As a consequence, they were 'jointly sued' by the New York State Income Tax Bureau for a large amount of money for back taxes which amounted to almost half a million dollars. Dreyfuss remembered at a meeting of the SID at Bedford Springs in 1953:

> At that time the three of us got together and hired some competent attorneys and we were the first group in New York who had ever been approached by the New York State Government to lick the situation and I can remember that we chose Walter, being the most professional of the three of us, to be tried first.[104]

Dreyfuss and Loewy would later jest that they 'paid these lawyers to make Walter a professional'.[105] Correspondence in Egmont Arens's archive suggests that Doris Marks Dreyfuss, Henry Dreyfuss's 'tax man',[106] secretary and wife, played an instrumental role in co-ordinating this effort, as she wrote to Arens in April 1940, advising him 'not to pay the tax' until after Teague's hearing.[107] The original case file (*Teague* v. *Graves*, 1941) makes the material motivation of professional status very clear. Teague's lawyer claimed in his defence:

> The graduates from Universities, institutes and schools who have scholastic degrees as Industrial Designers doubtless will be regarded as professional men. It is paradoxical that the petitioner and his present associates now engaged in the field, who are lecturing in these courses and teaching these students, should be classified as otherwise.[108]

This representation of educational stature was an exaggeration. As Harold Van Doren admitted in his career-guidance manual for the profession, published in 1940, opportunities for further education in industrial design were very limited.

In opposing the movement, the lawyer denounced the 'extreme liberalisation' of the term 'profession', which originally referred only to 'the three learned professions: divinity, medicine and the law'.[109] He stated:

> It seems to me that the petitioner's calling cannot possibly be regarded as the practice of a profession. He is neither an architect nor an engineer nor has he had any extensive course of specialized instruction and studying a field of science. A background of practical training and education does not itself raise the dignity of his occupation to that of a profession.[110]

Claiming tax under the category of 'profession' rather than 'occupation' therefore made a significant difference to income, a fact that is marked on the tax returns for industrial design consultant Arens. Arens was in regular contact with Dreyfuss's office regarding the status of unincorporated business tax. Following the success of Teague's claim in 1936, Arens wrote to the tax office to advise them of a change in the status of his profession. Whereas in 1934, he had filed tax claims under the 'occupation' of 'advertising artist' for clients including Calkins + Holder, by 1936, he was submitting claims under the title of 'Industrial Designer' for the same client, scoring out 'occupation' and writing 'profession' in its place, an act that could be viewed as poignant and expressive on the one hand and entirely pragmatic on the other.[111] For the SID, professionalism served this highly instrumentalized function, and Teague regularly referred to its utility on the basis of economic security and material privilege. This calculated view was not necessarily shared with his peers – Dreyfuss expressed loftier ideals on professionalism on many occasions – but Teague was always clear-eyed about its financial reward.

Gender infused early conversations about the definition of the industrial designer's role within both organizations, where the professional value of disciplines and practices was weighted according to gendered ideals of masculine work. In the US, designers expressed a distaste for the term 'stylist', which some rejected for its 'superficial, flighty' connotations. As one SID member put it, 'Industrial Stylist also seems to be gaining particular flavor; at least to style is a common verb; but the word is unfortunate because it smacks of millinery and suggests exactly the sort of superficial dressing-up job that a good designer does not do.'[112] The language of industrial design consultants, in defining their role against specialization, was highly charged with masculine associations of mastery, control and management. When explaining the distinction between the IDI and SID, Walter Dorwin Teague resorted to heavily gendered language: 'While membership of SID should have made it clear exactly what kind of designer you were', membership of the IDI merely indicated that you were 'presumably, some sort of a designer, whether textile, shoe, interior or possibly even industrial'.[113] The gendered overtones of this statement are bold and clear: textiles, shoe and interior design practices were widely understood to be firmly in the female realm. Nevertheless, even the ADI, which had textiles and fashion designers in its membership, was openly hostile to associations with fashion. In 1951, the newly renamed IDI established the first national IDI Design Award, publicly stating that the award 'could be given in any category of design except fashion'.[114] In 1963, a profile of Eliot Noyes in *Fortune* emphasized his disassociation from styling, quoting him as saying that 'a man who had tried to "glamorize" a washing machine, thus interfering with its usefulness, should be strung up by his thumbs'.[115] Peter Müller-Munk wrote to Dreyfuss to note how offended he had been by the article, and that he took particular issue with Teague being described as 'flamboyant', a descriptor he found unprofessional, presumably on the basis that it contravened the masculine self-image he held of the profession.[116]

Problems of a new profession

In spite of this public visibility, upon closer inspection, the field of industrial design held a relatively shallow basis for a claim to professional status in the US. In particular, as many commentators noted, the educational stature of the profession was underdeveloped, especially by comparison with many European countries. Key actors of influence remained unconvinced by the claims made by individual designers and their professional organizations, including one of the profession's greatest economic successes and maverick characters, Norman Bel Geddes (a 'decidedly medieval autocrat' who 'did not like any kind of organization').[117] MoMA in New York in particular played a somewhat critical role in providing a setting for an important conference, 'Industrial design: a new profession' in November 1946, organized and convened by Edgar Kaufmann, Jr, which put a spotlight on questions of self-definition, professional identity and the role to be played by professional organizations. Introducing the two-day event, Kaufmann explained the aims and purpose as follows:

> The SID kept telling us they had a really serious profession in hand ... We felt their strong statement that they were a profession was most interesting. It has been made formally and by implication many times before, by other design organizations and designers besides SID ... We therefore asked the Society of Industrial Designers if they did not believe the problem of a new profession with new problems to solve, new conditions to meet, could be discussed to mutual advantage by their own organization and representatives in other interested groups in the community, namely other design groups, other professional groups, such as architects and lawyers and perhaps most important, people interested in education.[118]

It is clear from this language that Kaufmann was keen to present the SID's role as one actor within a wider network of organizations and individuals, to dispel the view that this small group of design consultants should dominate. Dean Hudnut, from the Graduate School of Architecture at Harvard, was chairman and spoke optimistically in his opening remarks of the potential to raise the status of industrial design. Speakers from other professions – including a lawyer from General Motors – and educators were present to offer their definitions of a profession and their view of how industrial design fitted this model. Importantly, John Vassos, founder of the ADI, was there to give his own account of that Society's history:

> The ADI has existed since 1938 ... Like Mr Sakier, I am a pioneer in our profession and am glad that our Institute wants the young designers in America to get their place in the sun. We screen them carefully before they enter our organization. We have a code of ethics ... We in the ADI are anxious to have a profession. I hope this discussion at the MoMA will help to bring about our mutual objectives.[119]

Vassos's subtle attack at the SID here draws attention to perceptions of the SID's inflated self-importance as the 'pioneers' of the profession and the ADI's claims to represent young, as well as 'successful' designers. The ensuing discussion, which centred on key themes including status, licensing, responsibility and education, would form the basis for a recurrent set of problems for the profession. Much of this discussion revealed the self-aggrandising aims of the profession's chief protagonists, as they spoke with confidence on their privileged position as taste-leaders. Moving to address the question of social responsibility, Hudnut revealed the racial basis on which these perceptions of superiority were based:

> Speaking of this question of social responsibility to my wife, I said, 'we ought to start right here in our own house. Here we have a colored maid, named Nellie, and you give her the most horrible set of dinnerware to eat on (it had a pheasant and a rose and a castle). It is our job to teach her some good taste; it is our social responsibility'. I was so eloquent that my wife said 'yes, I will give her my best set. I will give her my Russel Wright'. And the next day Nellie gave notice.[120]

Hudnut's association between good taste and whiteness underlines, with startling clarity, the operation of racial ideology as an active feature of professional design discourse in these formative years. Russel Wright, a designer associated with the modernist mid-century American aesthetic, often 'interchanged' European and colonial motifs in his ceramics.[121] In colliding the concepts of good taste and social responsibility, this statement reveals the narrow and socially exclusive terms on which designers based their relationship with the public, a feature of the profession that would come under major critique, as we shall see in later chapters.

Observers to the discussion, including Hungarian émigré photographer and painter László Moholy Nagy, who established the Illinois Institute of Technology, was not a member of either organization and made some cutting summary remarks on their limited significance and impact to his work, stating that industrial design was 'just groping towards the profession … it is still mainly an adventure, a kind of goal'.[122] Others focused on the inadequacies of the US profession when compared with its European counterparts, where the issue of public responsibility, 'Good Design' and the education of taste (key goals of the British design reform movement), had played a more prominent role. Mr Bourdeau said:

> You cannot impose upon America the European philosophy of good any more than we can impose our philosophy on Europe. Our entire economy is based on accelerated obsolescence. That is the basis of our economy: accelerated obsolescence. It fits us perfectly, just as our type of government seems to fit us perfectly.[123]

These remarks further emphasised the spirit of independence and inward-looking gaze that characterized the pursuit of professionalism in US industrial design.

Transatlantic exchange

As design historian Robin Kinross stated in the *Journal of Design History* in 1990, the relationship between British and American design practice in the inter-war period has been frustratingly 'hard to pin down'.[124] Indeed, while there are many hints towards interaction, emulation and mimicry between the two professionalizing fields, in actual fact, little evidence exists to make this concrete, as both cultures of professionalism appear to have been centred around inward-looking ideals that reflected national values, ideas of cultural superiority and exceptionalism. At the first meeting of the SID in New York in 1944, Donald Deskey and Raymond Loewy reported:

> the SIA in England has asked for information concerning the SID and the Executive Secretary was authorized to extend a greeting on behalf of the Society to the English organization and to offer cooperation on general terms. The nature and extent of such cooperation were left for later decision.[125]

This relatively mild acknowledgement of the SIA's parallel objectives reflects the narrow and internally focused nature of professionalization for both organizations and also, possibly, their limited capacity to make a meaningful proposition of co-operation at this early stage. In fact, while both broadly sought to establish the identity of the designer as a professional, their organizations differed considerably in composition and outlook. The SIA was founded by a group of graphic artists and illustrators, who sought to co-ordinate the professionalization of industrial art as a field of specialisms. In the US, the professionalization of graphic design had taken place independently of industrial design.[126] The SIA had more in common with the organizational structure of the ADI, an organization founded to represent all 'specialisms' of industrial design on a broad level, than the SID, a narrow and exclusive society founded by and in the interests of Consultant Designers based in the New York.

Agency was distributed differently between the constitutive groups of professionalization in Britain and the US, which included the government, the press, business, professional organizations and designers and their offices. Whereas in Britain, professionalization was directed by a closed circle of 'design reformers', in the US, the impetus for professionalization came from designers themselves, from within the offices of design consultancies and through the media. Nevertheless, there were exceptions to these generalisations. By the post-war period, an individualized image of the designer had started to emerge in Britain through public exhibitions, including 'Britain Can Make It'. In the US, the public image of the industrial designer presented by the publicity office of the New York World's Fair worked to complement the individual design consultant and the corporation they worked for, proving the extent to which professionalization was entwined with cultures of consumption in both places.

For designers who moved between the two cultures, especially in the inter-war years, this cultural contrast acted as something of a stimulus, invigorating ideas about professional identity at home. Personal memory, recorded through letters in designer archives, offers revealing glimpses into the attitudes that framed design discourse in the two places. For instance, US packaging designer Walter Landor – who was born in Munich, Germany, and started his career in London, working with Milner Gray, Misha Black and FHK Henrion in the Industrial Design Partnership – described himself as

> frightfully British in my attitude and in my language ... I became adjusted to the English way of thinking and of relating to each other. I became imbued with understatement. One doesn't assert oneself too much and one doesn't brag. One is very subtle in one's assertions and so on. I had a wonderful time in England during that period. I grew tremendously.[127]

Writing to a colleague in later years, Landor remembered Britain as a 'beautiful challenging country, which taught me all that I ever learned'.[128] Landor was directly involved in the formative years of the profession in Britain, where he was a member of the advisory committee during the formation of the government-sponsored NRIAD.[129]

A more discernible dialogue between British and American design practice emerged after the war, particularly through the establishment of *Design* magazine, the 'mouthpiece' of the CoID and *Industrial Design* (*ID*), the first magazine of design criticism in the US. In her comparative history of these two texts, Alice Twemlow argues that they developed a 'shared language' after the war as their writers and art directors 'kept a sharp eye on one another'.[130] Moreover, as she points out, this relationship intensified in the 1950s, as British social and cultural critics 'were absorbed by American economic and cultural values'.[131] *Design*, its editors and staff writers, played an important role in steering British attitudes to professionalism in design, which were gentlemanly, dignified and defined in opposition to the more commercially oriented, 'buccaneering' persona of the US design pioneer. Jonathan Woodham points to an article in 1960 in particular, entitled 'Consumers in Danger', which described the exploitation of the consumer through manipulation techniques of the 'new professions' in the US.[132] Indeed, the representation of these practices as entirely American in origin and character went some way to distracting attention from their existence within many parts of the British design profession. Michael Farr took a particularly 'moralizing tone' in his reviews for *Design* magazine and in his book *Design: A Mid-Century Survey*, in which he warned against the 'wrong and self-assertive element in the American character'.[133] Farr takes great pleasure in reporting on a talk given by George A. Jergensen, head of Industrial Design at the Art Center School, Los Angeles, to a group of SIA members and CoID staff in January 1960:

> Misha Black thought he detected a character in the work which was different from other American schools. W. M. de Majo rephrased this by calling it frankly commercial. Jack Howe said it looked as though the Art School Center was concerned only with applied styling and that real design should grow from the inside – from a logical consideration of the basic requirements. Battle was joined and criticism came thick and fast. But it was a very one-sided battle for Mr Jergensen seemed unaware of his critics' meaning and one was left with the impression that industrial design in America is totally different in concept from its European equivalent.[134]

This idea of British cultural superiority was also reflected in tensions between the 'new' and the 'good' that shaped transatlantic design discourse and exchange.

The US preference for the 'new' positioned the designer as a protagonist for change, directed towards planned obsolescence, a governing ideology that was accepted, for the first thirty years of the profession at least, as an indicator of progress, in the same way as professionalization.[135] In September 1958, *Business Week* ran an article entitled 'The idea of perfectibility', in which they declared:

> nothing about America has so astonished other nations – or has been so frequently satirized by them – as our willingness to junk the almost new to adopt the brand new ... With other nations moving fast – not just the Russians but the Germans and many others – this is no time for Americans to slip from the creed that made this nation great – the creed that de Tocqueville called the 'idea of indefinite perfectibility'.[136]

Some of the most enthusiastic proponents of American obsolescence were those who had emigrated to the country between the wars. French-born Raymond Loewy, who accepted his American citizenship as one of the proudest achievements of his life, famously described the sales curve as 'the loveliest curve I know'.[137] Design practice was driven by a sense of mobility, which many leveraged to their economic success and professional acclaim in multiple locations. Loewy, for instance, had offices in New York and London from the inter-war years and his presence in both countries was a source of considerable self-esteem. Indeed, his offices in London, New York and later, Paris provided him with the opportunity to play multiple professional identities for different audiences. Throughout his career he performed the image of the professional and the commercial entrepreneur, a fine line that was expertly navigated by his publicist Betty Reese, as Chapter 3 will explore further. His performance of professionalism in both countries showed that they were not contradictory spaces, but relational. In the same way, British *Design* magazine situated its studied form of anti-commercial criticism in relation to the more overtly commercial discourse of US *Industrial Design*. The crossing of these currents of design discourse was formative to the development of a professional identity for the industrial designer in both places, an argument that is taken further in the next chapter.

Conclusion

Several themes and 'problems of the profession' raised in this chapter will recur throughout the analysis of this book, including the British denial of its roots to advertising and rejection of its status as a 'new profession' in favour of the traditional 'old professions' informed by the Victorian 'professional ideal' of gentlemanliness. In the US, divisions between the ADI and SID reflected a hierarchical and cultural division between the values of the 'in-house' designer and the design consultant; a rift that would widen over time. Broadly speaking, both cultures experienced the challenge of finding a place for industrial design that would fit within the ideals of the 'new' and 'old' professions; a problem that would recur and intensify in significance over time in both places.

Where the designer in the US was made visible through media representations in print, television and radio and autobiographies, the British designer remained 'anonymous', largely invisible and obscured by the hazier nomenclature of 'industrial artist'. This anonymity was partially shaped by a more cautious approach to professionalization on the part of design reformers and propagandists in Britain, who sought to promote *design* and not the *designer*. It might be said that while the US image of the industrial designer was shaped by a high degree of visibility, the identity of the British industrial artist was shaped by its absence. Designers were rarely featured in newspapers or magazines and the role of the industrial artist was barely understood by business people, never mind the general public. This invisibility was a source of concern and frustration for many designers and those involved in its promotion, as the designer was said to occupy an 'anonymous' position in Britain. By the post-war period, British design reformers had begun to pin their hopes on the role of the design consultant to bring greater visibility and recognition to the designer's status in industry. The pioneering role of the design consultant is frequently presented in design history as a manifestation of the overlapping aims and ideals of professionalization in both countries. Indeed, the image of the Consultant Designer was, perhaps above all, the most potent manifestation of the myths and ideals involved in the invention of the professional designer in the US and subsequently in Britain. The next chapter looks more closely at this role.

Notes

1. 'English Kids', *Life* (1943), pp. 87–90. Raymond Loewy Archive, CHMA.
2. Arthur J. Pulos, *American Design Ethic: A History of Industrial Design* (Cambridge, MA: MIT Press, 1965), p. 224.
3. The article does not state the nature of Loewy's friendship with Havinden, or indeed, how they were connected. However, Loewy opened an office in London in 1936, run by British designer Douglas Scott, and was well connected there.
4. Eric Hobsbawm and Terence Ranger, *The Invention of Tradition* (Cambridge: Cambridge University Press, 1983).
5. Geoffrey Millerson, *The Qualifying Associations: A Study in Professionalization* (London: Routledge, 1964), p. 159.
6. Heidi Egginton and Zoë Thomas, *Precarious Professionals: Gender, Identities and Social Change in Modern Britain* (London: University of London Press, 2021).
7. Jonathan M. Woodham, *Twentieth-Century Design* (Oxford: Oxford University Press, 1997), p. 167.
8. Commenting on the term Industrial Artist in *The Studio* in 1931, Frank Pick stated: 'Industrial Artist: the phrase is not a particularly happy one, but it must serve', press clippings, Box 98, CSDA.
9. See for instance Stana Nenadic, *Craftworkers in Nineteenth Century Scotland: Making and Adapting in an Industrial Age* (Edinburgh: Edinburgh University Press, 2022).
10. Wyndham Lewis (1929), quoted in D. L. LeMahieu, *A Culture for Democracy: Mass Communication and the Cultivated Mind in Britain Between the Wars* (Oxford: Clarendon Press, 1988), p. 154.
11. Helen McCarthy, 'Parties, Voluntary Associations and Democratic Politics in Inter-War Britain', *Historical Journal*, 50:4 (December 2007), 892.
12. D. L. LeMahieu, *A Culture for Democracy: Mass Communication and the Cultivated Mind in Britain Between the* Wars (Oxford: Clarendon Press, 1988).
13. 'DIA, Its History and Its Aims' (1929), Box 98, CSDA.
14. Herbert Simon, quoted in LeMahieu, *Culture for Democracy*, p. 158.
15. Noel Carrington, oral interview with Robert Wetmore (15 March 1984), CSDA.
16. Fiona MacCarthy, 'Demi-Gods or Boffins? The Designer's Image, 1936–1986', in *Eye for Industry* (London: RSA, 1986), pp. 17–22. See also MacCarthy, *All Things Bright and Beautiful: Design in Britain, 1830 to Today* (London: Allen & Unwin, 1972).
17. Signatories on the Article of Association: Septimus Scott, Harold Stabler, Lilian Hocknell, Oliver Bernard, James Norton (the only one to have identified himself as 'designer') and Milner Gray, SIA Articles of Association, Box 98, CSDA.
18. SIA Manifesto (1929), CSDA.
19. 'Men of Design', *News Chronicle*, Box 98, CSDA.
20. Interview dated 1937, unknown source, Milner Gray, RDI Boxes, FRDIA.
21. John Blake and Avril Blake, *DRU: The Practical Idealists* (London: Lund Humphries, 1964), p. 34.
22. David Preston, 'The Corporate Trailblazers', *Ultrabold*, Journal of St Bride Library (Summer 2011), 14.
23. SIA prospectus (1930), Box 98, CSDA.
24. Eric Fraser, 'The Bibbing of the Patrons at the Annual Dinner of the Bassett-Gray studio' (21 December 1929), Eric Fraser Archive, VAAD, AAD/2000/14/72.
25. See Leah Armstrong, 'Designing a Profession: The Structure, Identity and Organization of the Design Profession in Britain, 1930–2010' (PhD thesis, University of Brighton, UK, 2014).
26. Milner Gray, First Meeting of the North Staffordshire Group of the SIA (1931), Box 95, CSDA.
27. Cheryl Buckley, *Potters and Paintresses: Women Designers in the Pottery Industry, 1870–1955* (London: The Woman's Press, 1990).

28 Milner Gray, SIAD Presidential Address, 1968, 'The Price and the Value', Box 200, CSDA.
29 Ren Pepitone, 'Brother Barristers: Masculinity and the Culture of the Victorian Bar', in Heidi Egginton and Zoe Thomas (eds), *Precarious Professionals: Gender, Identities and Social Change in Modern Britain* (London: London University Press, 2021), p. 87.
30 Sir Henry McMahon, 'Institution of a Distinction for Designers for Industry', Dinner at the Society's House (13 November 1936), *RSA Journal* (November 1936), pp. 2–10.
31 Minutes of a meeting of RDIs to consider the formation of an association (9 November 1938), FRDIA, RSA/PR/DE/101/121/1.
32 Faculty of RDI minutes (2 February 1939), FRDIA.
33 Faculty of RDI minutes (31 January 1950), FRDIA.
34 In 1943 a 'designer for plastics' was sought for consideration. In 1948, 'the need for an appointment in dress design' was emphasized. Faculty of RDI minutes (12 May 1948), FRDIA.
35 Milner Gray, founder of the SIA, later stated that he had 'studied the older professions' before setting out to establish a professional organization for design in Milner Gray, 'The Beginnings of the SIA: Design in the Thirties', V&A Museum talk (8 November 1979), Box 98, CSDA.
36 Frank Pick, *The Studio* (1931), CSDA.
37 Milner Gray, 'The Beginnings of the SIA', Box 98, CSDA.
38 Misha Black, '1930–1955, A Quarter Century of Design', *SIA Journal*, Box 95, CSDA.
39 Milner Gray, interview with Bridget Wilkins, *Typographica*, 40:1 (Winter 1991), 6–8.
40 NRIAD, CoID First Annual Report (1945–1946), p. 17. Design Council Archive, UBDA.
41 Armstrong, 'Designing a Profession', p. 272.
42 I am grateful to Harriet Atkinson for pointing this out to me. See also Harriet Atkinson, *Modernist Exhibitions in Britain for Propaganda and Resistance, 1933–1953* (Manchester: Manchester University Press, 2023).
43 Noel Carrington, *Industrial Design in Britain* (London: Allen & Unwin, 1976), p. 154.
44 Wells Coates to Milner Gray (12 December 1945), Box 95, CSDA.
45 Peter Ray, 'SIAD: The First Forty Years', *The Designer* (October 1970), 4.
46 James de Holden-Stone, *Penrose Annual*, 45 (1951), 55–7, Box 207, CSDA.
47 Harriet Atkinson states that Round Boys was a colloquial phrase used to describe the architects who had been to architecture college, versus the 'flat boys'– graphic artists who had gone through the technical school system, *Festival of Britain: A Land and Its People* (London: I. B. Tauris, 2012), p. 41.
48 SIA Annual Report (1947), Box 95, CSDA.
49 Visiting figures were 1,432,546. Michael Farr, *Design in British Industry* (Cambridge: Cambridge University Press,1955), p. 234.
50 Atkinson, *Festival of Britain*, p. 44.
51 Catherine Moriarty, 'A Backroom Service? The Photographic Library of the Council of Industrial Design, 1945–1965, *Journal of Design History*, 13:1 (2000), 39–57.
52 Elizabeth Darling used the phrase 'fledgling designers' in 'Exhibiting Britain: Display and National Identity 1946–7', VADS learning module, VADS learning, www.vads.ac.uk/learning/designingbritain/html/esd_author.html, accessed 1 January 2015.
53 Lesley Whitworth, 'Britain Can Make It at the University of Brighton Design Archives', in Diane Bilbey (ed.), *Britain Can Make It: The 1946 Exhibition of Modern Design* (London: Paul Holberton/V&A Museum/University of Brighton Design Archives, 2019), p. 15.
54 Ibid.

55 See Robert Gordon-Fogelson, 'Vertical and Visual Integration at Container Corporation of America', *Journal of Design History*, 35:1 (March 2022), 70–85. John R. Blakinger, *Gyorgy Kepes: Undreaming the Bauhaus* (Cambridge, MA: MIT Press), pp. 27–9.
56 Norbert Dutton, quoted in Whitworth, 'Britain Can Make It', p. 201.
57 Misha Black, 'Birth of an egg cup' booklet (18 February 1947), Misha Black Egg Cup Folder, Design Council Archive, UBDA.
58 Dorothy Goslett, interview with Robert Wetmore (23 January 1983), CSDA.
59 Typescript of television programme for Granada TV, 'Saucepans and mammals' (20 June 1960). Misha Black Archive, VAAD.
60 Whitworth, 'Britain Can Make It', p. 15.
61 Mr Ironside to Mr Jarvis, with corrections to 'planned copy for "birth of an egg cup"'. Ironside made a correction to page 15, 'He uses decoration as a woman uses make-up': 'This may, I'm sorry to say, be objectionable to some teachers and should be avoided' (30 June 1947), Misha Black Egg Cup Folder, Design Council Archive, UBDA.
62 Jonathan Woodham, 'The Politics of Persuasion: State, Industry and Good Design at the Britain Can Make It Exhibition', in Patrick Maguire and Jonathan Woodham (eds), *Design and Cultural Politics in Postwar Britain: The* Britain Can Make It *Exhibition of 1946* (London: Leicester University Press, 1997), p. 59.
63 Darling, 'Exhibiting Britain'.
64 David Carter, interview with Robert Wetmore (19 March 1981), interview tapes, CSDA. Bell-bottom trousers are likely to refer to the style worn by members of the Royal Navy, since Carter had recently left the Navy after the war.
65 Report by Noel Carrington for CoID (5 December 1947), Britain Can Make It, Misha Black Designing an Egg Cup folder, Design Council Archive, UBDA.
66 See Meg Jacobs, *Pocketbook Politics: Economic Citizenship in Twentieth-Century America* (Princeton, NJ: Princeton University Press, 2005).
67 Meg Jacobs, 'Democracy's Third Estate, New Deal Politics and the Construction of a "Consuming Public"', *International Labour and Working Class History*, 55 (Spring 1999), 27. See also Roland Marchand, *Creating the Corporate Soul: The Rise of Public Relations and Corporate Imagery in American Big Business* (Berkeley: University of California Press, 2001).
68 Jacobs, 'Democracy's Third Estate', 27.
69 George Nelson, 'Both Fish and Fowl', *Fortune* (February 1934).
70 Roland Marchand, 'The Designers go the fair: Walter Dorwin Teague and the Professionalization of Corporate Industrial Exhibits, 1933–1940', *Design Issues*, 8:1 (Autumn 1991), 8.
71 C. Wright Mills, "The Man in the Middle", *Industrial Design*, November 1958, p.70. Obsolescence, IDSA_88_1, IDSA Archives, SCRC.
72 Arthur J. Pulos, *The American Design Adventure* (Cambridge, MA: MIT Press).
73 George Nelson, 'Both Fish and Fowl', *Fortune* (February 1934), 40.
74 Brooks Stevens, 'Planned Obsolescence in the Home Appliance', 7th Annual Appliance Technical Conference of the American Institute of Electrical Engineers, Wisconsin Hotel, Milwaukee, WI, 15 May 1956. IDSA 63_2, SCRC.
75 Gordon Lippincott, *Design for Business* (Chicago: Paul Theobald, 1947), p. 31.
76 Jan L. Logemann, *Engineered to Sell: European Émigrés and the Making of Consumer Capitalism* (Chicago: University of Chicago Press, 2019), p.5.
77 Ibid.
78 Henry Dreyfuss, 'Papers from the 1953 meeting of the American Association of Advertising Agencies', IDSA Archives, SCRC.
79 See Leah Armstrong, '"Fashions of the Future": Fashion, Gender and Industrial Design in US *Vogue*, February 1939', *Design Issues*, 37:3 (2021), 5–17.
80 Jacobs, 'Democracy's Third Estate', p. 27.

81 'Fashions of the Future', US *Vogue* (February, 1934), pp. 5–17.
82 The first Industrial Design training course was established at Carnegie Mellon Institute of Technology in 1934. In 1954, Sally Swing, secretary at the SID, estimated that approximately '200 students complete training in the field each year'. Sally Swing, 'Industrial Design: A Profession', *Career News*, 13:1 (February 1954), n.p., IDSA Archives, SCRC.
83 Nelson, 'Both Fish and Fowl'.
84 Donald Deskey, 'Designing Men', US *Vogue* (February 1939), p. 72.
85 Raymond Loewy, 'Fashions of the Future', US *Vogue* (February 1939), p. 141.
86 See Gilbert Rohde, 'Fashions of the Future', US *Vogue* (February 1939), pp. 138–9.
87 See Joanna Bourke, 'The Great Male Renunciation: Men's Dress Reform in Inter-War Britain', *Journal of Design History*, 9:1 (1996), 23–33.
88 In her biographical history of Rohde, Phyliss Ross offers a different interpretation of his contribution, attributing it as a publicity exercise, in an effort to stand out from the other designer's submissions. Phyllis Ross, *Gilbert Rohde: Modern Design for Modern Living* (New Haven, CT: Yale University Press, 2009).
89 Rohde, 'Fashions of the Future', pp. 138–9.
90 Ibid.
91 Edgar Kaufmann, Director of Department of Industrial Design, MoMA, Industrial Design Standards, England's Council of Industrial Design, Henry Dreyfuss Archive, H.TS171.4D7D1954, CHMRB.
92 Ray Spilman, interviewed by Arthur J. Pulos for SID Archive, c.1968. Transcript, Box 37, Folder 2,VAAD. Arthur J. Pulos Archive, AAA.
93 ADI membership roster, 1944–1945. IDSA Archives, Box 65, SCRC.
94 John Vassos to Norman Strouse (5 July 1962), John Vassos papers, Box 1, Correspondence, IDSA Archives, SCRC.
95 Ann Franke discusses education; Belle Kogan as chairman discusses steps taken to license the profession of industrial design in New York and Ruth Gerth chairs a committee on competition work. Lilian Hollander was 'busy with war work and looking forward to going back to her interesting work of designing silver'. *ADI Newsletter* (1945), Leon Gordon Miller papers, SCRC.
96 John Blake, *Design* (1963), cited in a letter from George Beck to John Vassos (30 September 1963), John Vassos papers, Box 1, Correspondence, SCRC.
97 In a letter to Dreyfuss in 1965, he explained, 'Alexander Kostellow, Belle Kogan, Ann Franke and I went to Albany for a preliminary hearing on this subject. The result was that we were advised that both societies sponsor such a request and ASID declined. So we dropped it.' John Vassos to Henry Dreyfuss (3 September 1965), Vassos papers, Box 1, Correspondence, SCRC.
98 'My guess is that we will be able to pull out of the ADI all of the best Industrial Designers. We can get more by cooperating with this group than by fighting it', Egmont Arens to Philip McConnell (26 March 1945), IDSA papers, Box 64, SCRC.
99 Arthur BecVar to Ray Spilman (20 June 1962), IDSA Records, Box 65, SCRC.
100 Raymond Spilman, interview with Arthur J. Pulos, c.1968, Pulos papers, AAA.
101 Ibid.
102 Board of Directors Meeting (14–15 October 1948), IDSA 2, IDSA papers, SCRC.
103 In April 1940, Arens received a letter from Doris Marks, Dreyfuss's wife, secretary (and self-proclaimed 'tax man') advising him not to pay the unincorporated tax bill he had received until they had heard the results from the Teagues' hearing, which they all hoped would 'straighten us out', letter from Doris Marks, Henry Dreyfuss's Secretary (16 April 1940), Egmont Arens papers, Box 64, SCRC.
104 Henry Dreyfuss, '1953 Design Conference, Sponsored by the Society of Industrial Designers', Bedford Springs, Pa. (16–18 October 1953), Arthur J. Pulos Archive, Box 58, AAA.
105 Ibid.

106 Doris Marks Dreyfuss to Egmont Arens (16 April 1940), Egmont Arens papers, SCRC.
107 'We are waiting notice from Albany for a hearing. I believe Teague's hearing is the first one to come up according to schedule'. She advises him to prepare for a long battle, 'unless Teague's case straightens us out (which we all hope it will), our next hearing will be more fight'. Ibid.
108 Photocopied case notes from *Teague* v. *Graves* re: taxation of the professional designer from New York, supplement, 2nd series, p. 766, Arthur J. Pulos Archive, Box 37, folder 2, AAA.
109 Ibid.
110 Ibid.
111 Egmont Arens to New York tax office (14 April 1934–1936), Arens 64, SCRC.
112 Harold Van Doren, *Industrial Design: A Practical Guide* (New York: McGraw Hill, 1940), p. 15.
113 Walter Dorwin Teague to Peter Müller Munk (16 June 1955), Leon Gordon Miller papers, Box 1, ASID/IDI Correspondence, SCRC.
114 IDI, First National Design Award, IDI folder, IDSA, Box 65, SCRC.
115 'An Industrial Designer with a Conspicuous Conscience', *Fortune*, 1963, quoted in Noyes's obituary, 'Opposed Glamorization', *New York Times* (19 July 1977), p. 38.
116 Peter Müller-Munk (16 August 1963), Grievances folder, IDSA Archive, SCRC.
117 Arthur J. Pulos, Folder 2, Box 37, Annotated transcript for *The American Design Adventure*, Prologue. Arthur J. Pulos Archive, AAA.
118 Edgar Kauffman, Jr, 'Industrial Design: a New Profession', MoMA (11 November 1946). Henry Dreyfuss Archive, D.TS171.4D741954, CHMRB.
119 John Vassos, 'Industrial Design: A New Profession', MoMA (11 November 1946), CHMA.
120 Dean Hudnut, 'Industrial Design: A New Profession', MoMA (11 November 1946), CHMA.
121 Kristina Wilson, *Midcentury Modernism and the American Body: Race, Gender and the Politics of Power in Design* (Princeton, NJ: Princeton University Press, 2021).
122 Moholy Nagy, 'Industrial Design: A New Profession', MoMA (14 November 1946), CHMA.
123 James Bourdreau, 'Industrial Design: A New Profession', MoMA (13 November 1946), CHMA.
124 Robin Kinross, 'Émigré Graphic Designers in Britain: Around the Second World War and Afterwards', *Journal of Design History*, 3:1 (1990), 35.
125 First Annual Meeting, SID (17 December 1945), IDSA_2_1, Minutes 1945, IDSA Archive, SCRC.
126 See Eleanor Mazur Thomson, *The Origins of Graphic Design in America, 1870–1920* (New Haven, CT: Yale University Press, 1997).
127 Walter Landor, Reminiscences, Series 17, Box 17.1 personal papers, Walter Landor/Landor Associates Collection, NMAH.
128 Walter Landor to Greg Vallance, n.d., Bassett Gray folder, Walter Landor/Landor Associates Collection, NMAH.
129 During this time, Landor had a monthly column in *British Plastics* magazine, 'advocating for better design for industry', and was chairman of the Plastic Design Division of the Society of Industrial Artists.
130 Alice Twemlow, *Sifting the Trash: A History of Design Criticism* (Cambridge, MA: MIT Press, 2017), p. 12.
131 Ibid.
132 Jonathan Woodham, quoting Farr in 'Putting the industrial into design', in Maguire and Woodham, *Design and Cultural Politics in Post-War Britain*, p. 126.
133 Ibid. Twemlow also notes that Farr's outlook showed the connections between design reform and social reform; see Twemlow, *Sifting the Trash*, p. 31.

134 'Views on Design Training', *Design* (January 1960), 27.
135 Aarre K .Lahti, Professor of Industrial Design, University of Michigan, 'The Scope of Industrial Design', IDI Date Line, December 1962, 3:9, IDSA Archive, SCRC.
136 'The idea of perfectibility', *Business Week* (27 September 1958), p. 184. IDSA, Obsolescence Folder 88_1, IDSA Archives, SCRC.
137 When interviewed by the Chartered Society of Designers in Britain, around 1984, Loewy toned down this assessment somewhat, describing his approach to marketing as 'responsible'. Interview Tapes, CSDA.

2
The (General) Consultant Designer

'First comes the pioneer' in the 'normal evolution of a profession', said editor of the British Journal *Art and Industry* Frank Mercer as he addressed an audience of industrialists and designers at the Royal Society of Arts (RSA), London in 1945.[1] The purpose of Mercer's paper was to highlight the absence of an individual in the British design industry who could assume this role, pointing directly to the importation of the American model of design consultant as a solution to this problem. As we saw in Chapter 1, vibrant publicity in the US provided a bold and dramatic entry for this professional to public life, led by the 'pioneer' model. Indeed, in many cases, the term 'pioneer' and 'Consultant Designer' were used interchangeably, so that the Consultant Designer's status as the apotheosis of professionalism in design was assured and complete. Subsequent historicization absorbed this view of the consultant's pioneering role. As British design historian Penny Sparke put it in her history, *Consultant Design*, 'before that moment the designer had not fully realized his special position in modern culture; he had instead, leant on the skills of others, whether the fine artist, the architect, the craftsman, the engineer or the technician'.[2] While the structure, organization and identity of the industrial design professions in Britain and the US responded to and were shaped by national social, political and economic contexts, they both found their common ideal in the figure of a white, male design consultant, an enduring image that 'defined' the image of the designer for subsequent generations, and, arguably, sustains the profession into the present day. This chapter interrogates the significance of 'pioneer' model in the professionalization of industrial design and critically examines the cultures and ideologies that underpinned their 'special position' within the profession.

While the term 'design consultant' might be used somewhat ahistorically to refer to the employment of a designer to work for industry in an advisory capacity, the term also refers more specifically to the identity of 'the Consultant Designer for industry', as a product of colliding ideals between US corporate culture and industrial design in the US before the Second World War. This chapter explores the dynamics of these ideals, finding that they were informed by ideas of exclusivity, male privilege and a heavily mythologized agency of the individual in both places. Importantly, it was not a skill-based role informed or founded upon substantive technical or disciplinary-based skill, but rather a *title* invented to propel the designer to the upper echelons of industry and make the profession visible to the public. In self-defining the qualities of the Consultant Designer, its New York- based practitioners emphasized the qualities of good taste, judgement and 'generalist' expertise; qualities not traditionally considered requisite or indeed sufficient for professional status. In this sense, the Consultant Designer title was a kind of 'packaging' – or 'sociological wrapping' – around the professional identity of the industrial designer, without adhering to any of the structural qualifications of professionalism.[3]

The image of the Consultant Designer seemed to put the ideals of professionalism in Britain and the US in contact for the first time, through the adoption of the title 'General Consultant Designer' by the British Society of Industrial Artists (BSIA) in 1953. Indeed, the identity of the General Consultant Designer has no history in Britain without its relationship to the US profession, and historians have cited the 'adoption' of this role in Britain as evidence of the 'acceptance' of US business values in British industry.[4] While this chapter sketches out some shared characteristics between the two identities, it also tests assumptions about the 'importation' of this role from the US to Britain, finding that it had limited hold over the British design profession and limited acceptance. On the one hand, this can be explained by the fact that British attitudes to professionalization were driven by an adherence to specialization, bringing it into conflict with the generalist model of expertise presented by industrial design consultants in the US. It can also be explained by the relatively late adoption of the title, by which time US industry had already shifted towards 'integration' as the dominant method of employment. Meanwhile, in Britain, the emergence of a 'new wave' of consultancies identifying as 'European' in scope adopted an integrated model of marketing, advertising and design practices, moving beyond the individualized model of Consultant Designer, and responded to an increasingly open, international market. Therefore, while historians have tended to present the trend towards consultancy work in Britain as a simple absorption of US-style working practices, the chapter ends by reflecting on some of the unique features of these consultancies that locate them within a broader trend towards globalization in an open-market economy.

Management consultancy

In 1930, *Business Week* dramatically announced the emergence of another new professional service: management consulting, 'the world's newest profession'.[5] Management consultancy, a service with a scientific method that had been growing in the US and Britain since the mid-nineteenth century, rose to new levels of public visibility in the inter-war years. The post-war period saw the exponential growth of the sector in the US, which, by the middle of the twentieth century, had raised the status and identity of the 'consultant' to higher levels than those working within concrete specialisms. The consultant can be generically defined as a 'specialist provider in support of management in exchange for a fee', but like most professional roles, the limits and boundaries of this are in a constant state of negotiation and redefinition.[6] The history of the management consultancy profession is also dominated by pioneers, namely Frederick Winslow Taylor, whose scientific principles and practices, responding to the growth of firms and the complexities of new technologies, inspired the 'efficiency movement'. Taylor, an engineer, has been widely accredited with spearheading management consultancy as a distinctive form of expertise that could grow business.[7]

The management consultant was originally identified as independent, with objectivity and outsider status being considered crucial to their service. Impartiality and 'honesty' were critical, 'to the extent that he or she is free from the internal political wranglings of the firm'.[8] As Samuel Haber put it in his sociological history of professional culture in America, 'it was the full-time consultants who were the principal advocates of professional ideals among the engineers', and this was true across all branches of the discipline.[9] Haber explains that consultants, because of their similarities to the self-employed 'freelancer', were the loudest and most visible champions of professional status. While the origins of management consultancy was connected to scientific method and was advanced through a collaboration between the professions of engineering, accountancy and law, 'the growth of the management sector emerged as a profession to give men more career opportunities in the new employment hierarchy'.[10] The Consultant Designer imported this highly gendered discourse of expertise into the industrial design profession.

As cultural historian Jan Logemann states, 'In general, consulting firms had become a central feature of the American economy during the middle decades of the twentieth century' and after the war, they became influential factors in the transatlantic transmission of 'American' management concepts and practices.[11] Nevertheless, there were key differences in the trajectory of management consultancy in Britain and the US. While the inter-war period was an important time of consolidation and acceleration of management consultancy firms in the US, in Britain, there was a 'broad continuation of paternalistic approach adopted by employers of outside intervention in the firm, specifically with regard to the strategic decision making process', contributing to the significantly slower growth there.[12] This broader history forms a useful context from which to read the identity of the consultant designer in the analysis that follows.

The Consultant Designer in the US

The role of industrial design consultant received a highly flattering introduction to US business through the pages of *Fortune* magazine in February 1934, in an article anonymously penned by consultant and architecture and design critic George Nelson, which quickly entered the mythology of the profession.[13] While he would later be one of its most cutting critics, Nelson was a valuable proponent of the industrial designer in the formative years of the profession, with influential connections to the fields of architecture and design criticism.[14] His anonymity in penning the piece was important, as he adopted the tone of a detached observer, suggesting the 'omniscience and omnipotence' of this new profession.[15] In brief, the article presented the work of ten men – Donald Dohner, Henry Dreyfuss, Norman Bel Geddes, Laurelle Guild, George Jensen, Raymond Loewy, Georg Sakier, Walter Dorwin Teague, Harold Van Doren and John Vassos – all based in New York, all male, and said to be 'illustrative' and 'typical' of the new profession.[16] In putting the personalities of individual Consultant Designers at its centre, the article emphasized the values of individualism central to the professionalization of industrial design in the US. Nelson took particular interest in the eccentricities of the men, stating that they were 'group in occupation only', embedding the identity of the consultant within the cultural narrative of individualism. This included a lengthy section on Norman Bel Geddes's single-handed contribution to the US economy ('more than a billion dollars') through his commitment to obsolescence. There was 'handsome Harold Van Doren', 'bomb-thrower Geddes', 'suave, able, successful Walter Dorwin Teague'.[17]

The identity of the Consultant Designer's pioneering persona was the principal way Nelson sought to sell the 'seriousness' of this new profession to the *Fortune* reader. To do this, he presented the business credentials of the design consultant in tabular format, stating that they were usually a 'small group, hardly more than twenty-five in number, operating either alone or with small staffs. They have been specializing in their work for from four to ten years [sic]; have redesigned products in industries with a normal annual volume of over seven billion dollars; have earned themselves up to $150 000 apiece each year.'[18] The figures in 'Comparisons of a Few Leading Industrial Designers' (see Table 1) provided a flattering overview of the range of age, experience, salary, organization and scale of the industrial designer's employment.

Table 1 Comparisons of a few leading designers (1943)

	Age	Years as industrial designer	Previous experience	Compensation (royalties are subject to special arrangements)	Staff members	Typical achievements (starred item is work in progress)	Client manufacturer
Dohner	41	7	University design teacher Freelance designer	Cost of design department: $75,000 per year	8	Vacuum cleaner Mechanical water cooler Air-conditioning units	Westinghouse
Dreyfuss	29	5	Theatre sets	Flat fee: $1,000 to $25,000. Hourly consultation: $50	5	Washing machine Alarm clocks Check protector	Sears, Roebuck Western Clock Co. Todd Co.
Geddes	40	7	Theatre sets and costumes	Flat fee: $1,000 to $100,000. Royalties	30	Gas range Telephone index Radio	Standard Gas Equipment Bates Mfg Co. Philco
Guild	35	10	Art director Furniture expert	Retainer fee up to $25,000. Fee per day: $100 to $200. Flat fee $300 to $25,000. Royalties	4	Refrigerator; cooking utensils 'stoves to roller skates'	Norge Corp. Wear-Ever Aluminum Montgomery Ward
Jensen	35	6	Artist	Retainer fee: $500 to $20,000	3	Telephone Metal kitchen sink Water heater	AT&T International Nickel Co. L.O. Koven & Bro.
Loewy	40	6	Electrical engineer Art director	Retainer fee: $10,000 to $60,000. Flat fee: $3,000 upwards. Royalties	1	Motor car Duplicator Kitchen sink and bathroom units	Hupp Motor Car Corp. Gestetner Co. (British) Sears, Roebuck
Sakier	36	11	Mechanical engineer Art director	$15,000 to $25,000 income from design work. Retainers	11	Baths, washbasins, etc. Bathroom units Vacuum equipment	American Radiator & Standard Sanitary Accessories Co. Schellwood-Johnson Co.
Teague	48	6	Advertising designer	Retainer fee: $12,000 to $24,000. Flat fees: $500 to $10,000	4	Cameras Furnace Mimeograph	Eastman Kodak National Radiator A.B. Dick Co.
Van Doren	38	4	Painter Ghost writer	Consultation fee: $100 per day. Jobs executed: $500 to $5,000	8	Scales Kitchen grill Paint gun	Toledo Scale Co. Swartzbaugh Mfg Co. DeVilbiss
Vassos	35	7	Advertising agency illustrator	Retainer fee: $12,000. Flat fee: $1,000 to $7,000	3	Drink dispenser Turnstile Radios	Coca-Cola Perey Mfg Co. RCA

The article underscored the elevated position the Consultant Designer sought within the hierarchy of the profession. If the early years of the profession had been marked by 'pretenders and visionaries'; these Consultant Designers were the 'serious' professionals.[19] The gravity of their professional value was underscored by their association with major corporations, including Westinghouse; Sears, Roebuck; Philco; Coca-Cola and RCA, their fees and, as shown in another table, the profit margins generated. As would become clear, the identity – and trajectory – of the corporation and the Consultant Designer were closely entwined.

The factual basis of Nelson's research is hard to establish and likely to have been exaggerated. The archives of individual consultant office files give some insight into the wage brackets and employment conditions of those working within the 'big-name' consultancies in New York at the time. In the early 1940s, Egmont Arens, Raymond Loewy and Brooks Stevens exchanged scales of fees and job classification structures. According to Arens's notes, wage brackets were split into five categories – Apprentice; Junior; Senior; Job Captain; Department Head – with a scale from $20 to $150 per week. Secretaries and Receptionists were the lowest paid within the wage scale, alongside the Laboratory Assistant.[20] Loewy's office structure was more elaborate and notable for the sizeable publicity department, charged with 'preparation of articles for publication, interviewing publishers' representatives, obtaining information and photographs for news releases and retaining file of information and clippings'.[21] It included a Senior Architect, Draftsman, Designers in Training, File Clerks, Office Boys, Porter, Engineer, Stockroom Clerk, Project Director, Charwoman, Branch Office Manager, Assistant and Branch Office Manager. Stevens's office included a 'tracer; junior detailer; major or full-size layout man; junior checker; senior checker; group leader; junior and senior product illustrator; sketching artist and junior and senior stylist'.[22] The highly gendered language used to describe the office structure gives an insight into the masculine culture that governed these workspaces.

These classification structures also give some insight into the hierarchical culture that was being built up around New York-based design consultancies, informing attitudes to work within the profession. Speaking to 'art-school students' in the mid-1950s, Consultant Designer Russel Wright told students that the role of 'renderer' was the 'best meal ticket' in the design consultancy office.[23] The role could 'command a high salary' and 'when he [sic] can't find work in industrial design, he can work as an illustrator for advertising offices or magazines', highlighting hierarchical attitudes to professional status. Nevertheless, he said, they should be aware that 'a renderer seldom becomes a designer', and finding the right position was in some cases a question of temperament; 'an industrial design office cannot be a collection of erratic, long-haired, temperamental artists'. Thus, a clear hierarchy of status between the consultant and staff designer was presented, as he advised those who had been fired from a position in industrial design, or could not find work, not to go freelance, stating that it would be 'more sensible to become a specialist. Give up your idea of being an industrial designer, working on all types of products, putting all of your efforts in this one direction ... work as a staff designer in a factory.'[24] It is clear that he saw the role of the consultant as accessible to a privileged minority.

Consultant versus company man

By mid-century, the industrial design profession in the US could be divided into two groups; consultants with 'household-name' status and the 'unsung company man', who worked for corporations or manufacturing companies as salaried employers.[25] This status was, to a very significant extent, shaped by the emergence of publicity as a tool in the Consultant Designer's professional equipment. As an article in *Industrial Design* magazine put it, 'The company designer can turn to his company PR staff for help with a speech and will find the PR men eager to hear his explanation of the merits of a new product, but he will not be able to persuade the PR people to single out either his own activity or that of his department in their releases.'[26] By contrast, Consultant Designers invested considerably in the employment of an in-house or freelance publicist. The 'staff designer' was at a major disadvantage here, being effectively hidden and obscured from view in the shadow of the design consultant, a position that continued to divide the profession between the 'visible' and 'invisible'. Walter Dorwin Teague and others frequently defended their use of publicity, arguing that they needed it to attract clients, where the staff designer could be comfortable in the terms of their contract.[27] This distinction would sow the seeds of division within the US profession, which will be explored in Chapter 5.

In a 1947 Society of Industrial Designers (SID) publication entitled *Good Design is Your Business*, Richard Marsh Bennett defined the 'contemporary industrial designer' as one who 'takes the entire scope of machine design by virtue of his broad knowledge and ability to integrate the specialized work of others'. 'Unlike medicine,' he explained, 'the highest rewards are not for the designer who is a specialist but for one whose scope of knowledge embraces the entire productive and distributive process.' He went on:

> This is the designer who has captured the imagination of the public and the confidence of the industrialist ... He is the contemporary name designer, a professional man operating independently, backed by an expert organization combining many talents and contemporary disciplines – the industrialized industrial designer.[28]

Marsh Bennett was exaggerating when he said that the identity had 'captured the imagination of the public', a statement grounded in ambition and mythology. As Gordon Lippincott put it in a separate publication in the same year, the 'industrial designer' had been 'accepted as representative of a new profession, even though the majority of Americans have never heard the term, much less know what it means'.[29] Revealingly, Marsh Bennett located the expertise to the organization for whom the consultant worked, emphasizing the consultant's dependence on the corporation for professional authority. His statement makes explicit the link between professionalization and industrialization. Nevertheless, while industrialization was most commonly associated with specialization, as it had been in the British design context, the US conception of the Consultant Designer was modelled on the identity of the management consultant and the value of general over specialized expertise.

Designer and historian Arthur J. Pulos later explained that Americans

> revere the accomplishments of the generalists of the 18th century, remembering that men like Franklin, Jefferson and Whitney among others, ranged at will across all of the arts and sciences of their time. They know that man's mastery over the laws of nature in the 20th century has made it possible for creative and courageous men to become such generalists again. Industrial Design is one of the generalist professions that will lead this society into the 21st Century.[30]

In connecting the themes of masculinity, creativity and leadership, Pulos was locating the design consultant in a much longer cultural history of heroic masculinity. Russel Wright presented the fluidity of design practice in its early years as a source of great freedom and pleasure, recounting how 'we jumped blithely from one type of manufactured product to another, from cigarette lighters to automobiles, from vacuum cleaners to trains, glassware, comptometers and linotype machines'.[31] One of Raymond Loewy's favourite promotional quips was that he could design anything from 'lipsticks to locomotives'; Walter Dorwin Teague, 'matchstick to a city'.[32] Consultant designers frequently struggled to articulate their expertise, which was another reason many reverted to their publicists to do this job for them.[33] In June 1962, Arthur BecVar, consultant to General Electric, sent Ray Spilman a copy of a 'Master's thesis done in their office on management consulting', which listed nine desirable, basic traits of consultants: 'high level of intelligence; sensitivity to people and situations; self-sufficiency; basic professional skill; ambition; personal attractiveness to others; flexibility; energy; growth potential'.[34] This list reveals the personal, qualitative nature of the skills, with a surprising absence of trained expertise or education. In his career-guidance manual for industrial design, Arthur J. Pulos placed the Consultant Designer's role firmly in this area when he said, 'Designers are expected, above all else, to be men and women of good taste and sound aesthetic judgement.' Building on the theme, he advised young readers to ask themselves the question, 'Do you have good taste in clothing and in other personal products which you select for yourself?', to ascertain their suitability for the role.[35] As Chapter 1 argued, this can be partly explained by the absence of a well-established training system for the industrial designer in the early years of the profession, but it also reveals the highly performative nature of the role, which depended upon personality attributes, rather than skill or trained expertise.

The minority status of the Consultant Designer was put in perspective in a 1947 report by the American Management Association (AMA) in co-operation with the SID.[36] The AMA conducted a questionnaire-based study entitled 'American Business and Industrial Design', with the object of establishing how and where the designer was employed in US business. They received replies from 133 companies covering consumer goods, industrial goods, packaging and merchandising sectors. The report established that design decisions were taken at management level and that the use of an in-house staff designer was more common than the employment of a consultant in most sectors.[37]

In April 1954 the Eastman Kodak Company told the SID that it spent just 2 per cent of its design budget on consultation.[38] It is therefore clear that there was a disjuncture between the idealized position of the Consultant Designer, enjoyed by those at its pinnacle, and those positioned inside industry, the 'corporate' designers, the invisible majority

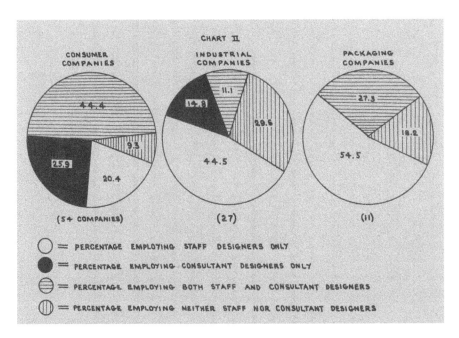

2.1 Charts from American Management Association for the Society of Industrial Designers Report (1947). Arthus J. Pulos Papers, Box 41, Archives of American Art, Smithsonian Institution, Washington, DC.

revealingly referred to as 'captive designers' – or 'staff designers'. From this perspective, it is fascinating to note how a small minority of the profession continued to take ownership for and express authority within the profession, boosted by the role of publicity, another 'new profession' more closely examined in the next chapter.

Cultural intermediaries

The Consultant Designer's role was mutually implicated in the ambiguous act of producing and selling, designating the designer's work as a 'cultural intermediary'.[39] This form of work, also carried out by publicists, advertising agents and marketing practitioners, can be characterized by the blurring of boundaries between 'personal taste and professional judgement'.[40] As Roland Marchand has convincingly argued, the Consultant Designer became the puppet of the corporation, precariously balanced between the act of making and selling consumer goods and 'softening' the self-image of corporate capitalism in the eyes of the general public. In this way, the 'young professions' would mediate the distance between business and consumer, corporation and the public. Thus, 'designers' new indispensable form of expertise' had 'little to do with aesthetic judgements, as traditionally understood', but could be better contextualized in relation to the rise of marketing and audience research.[41] 'Manipulation' and corporate image management were key skills of these intermediary professions and underpinned their position as interlocutors in the new economy. An early document written by Philip McConnell, Secretary at the SID, described the industrial designer as a 'creative interpreter who acts as a liaison between the consumer and the manufacturer'.[42]

Many designers enthusiastically took up this role, accepting their position as protagonists of consumer capitalism. This can perhaps be most blatantly seen in the appearance of industrial design consultants in public advertisements for industrial products, a practice that was most prolific in the 1950s in the US. For instance 'noted' package designers Walter Landor and Donald Deskey featured in an advertising campaign for Olin Cellophane in 1952 (see Colour Plates 3 and 4).

These two images visually convey the position of the design consultant as a mediator between industry and consumer, as the material transparency of the cellophane is used suggestively to communicate the value of the design consultants' opinion as trusted taste-leaders on matters of consumer choice. This can be seen particularly in Landor's advertisement, with the slogan 'Olin Cellophane Sells the Truth'. In another example from a different campaign, 'famous designer' Russel Wright (see Colour Plate 5) featured in a campaign to promote Du Pont 'Fabrilite' plastic fabric coverings. To see Consultant Designers participate in this form of blatant advertising presents a great irony to the qualities of professional objectivity, independence and 'freedom of mind and intelligence' many claimed to possess.[43] Pictured 'in his design studio', these examples, of which there are many, present a striking image of the close relationship between publicity, professionalization and consumer culture. This was a relationship that would soon undergo vigorous criticism from sociologists and cultural critics outside the profession, undoing some of the gloss and glamour of the 'pioneering' years of industrial design. While it would later be looked upon as a dubiously duplicitous activity from the perspective of professional ethics, many designers also later recognized this era as a high point in their professional self-image and influence.

Design consultancy in Britain

By the post-war period, British design reformers, including propagandist John Gloag, came to see the Consultant Designer as the missing ingredient in the professionalization of industrial design at home. Writing in his 1944 publication, *The Missing Technician*, Gloag praised the 'great progress of American industry' and argued that 'America's capacity for giving jobs to the right men' should be instructive for Britain.[44] British perceptions of inferiority and invisibility focused on the perceived absence of the consultancy role, which had become synonymous with status and independence – attributes British designers felt were severely lacking at home. Fuelled in part by cultural frustration and perceived inferiority at the low status of the British designer in industry, and with an eye on the well-publicised economic successes of the US Consultant Designer, sections of 'design reform' propagandists now sought to reorientate the identity of the industrial designer as a specialist to that of a General Consultant. F. A. Mercer's paper, 'The Industrial Design Consultant', quoted at the opening of this chapter, sought to convince his audience of manufacturers and influential men of business of the existence of this 'new and all-important profession', adopting a tone not dissimilar in style to Nelson's 1934 *Fortune* editorial. However, in stark contrast to Nelson's writing, Mercer's paper was filled with a sense of British inferiority and its enthusiasm was directed entirely at the professional status of the industrial designer in the US, seen to be 'far in advance of this country'. He praised the 'pioneering' efforts of designers Raymond Loewy, Walter Dorwin Teague, Norman Bel Geddes, Henry Dreyfuss, Egmont Arens and others, who were 'almost household words before the present war. Their prestige as industrial design consultants extended throughout the Union and far beyond it.'[45]

Mercer emphasized their breadth of scope, echoing the ambitious and bold language of US commentators: 'from the steamship, airplane or locomotive to the hairgrip, the compact or the man's pipe. Nothing is too great or too small to come within the province of industrial design and the industrial designer.'[46] In several places, the report used language that seems to have been plucked from the pages of US promotional literature. Mercer acknowledged at several points that he had consulted Loewy in the preparation for the paper and quoted with great admiration the scale of Loewy's successes in stores and architecture, products, transportation and packaging, emphasising to the British audience the great versatility that underpinned the role of the consultant in the US. Loewy, who established a London office in the mid-1930s, had recently been awarded the status of Hon. RDI (Royal Designer for Industry) within the Royal Society of Arts (RSA), an award in which he professed great pride, second only to his American citizenship.[47] Mercer continued to define industrial design as a specialist activity, but that of 'a new kind of specialist', to be used by industry in a new, more complete way, involved in every step of the design process, from planning to execution and even marketing, constituting a more holistic reading of specialist expertise. Responding to his comments, manufacturers in the audience met Mercer's enthusiasm with well-worn scepticism. A Mr T. S. Innes challenged the idea of consultant status, arguing for the importance of teamwork across specialist expertise, in the traditional form of industrialized professions. Another businessman, Mr Edward Richardson, who claimed to have employed Consultant Designers in the past, took issue on the basis of personality, stating: 'they want to do too much ... these big men in the designing world want to take on the commercial and selling side too', a statement that attests to the continued resistance to commercialism within art and industry in Britain and the unsympathetic relations between manufacturer and designer in Britain at this time.[48] Where in the US corporations and industry had come to see the value of the designer as a mediator between production and consumption, placing the Consultant Designer in this position, this view had not yet taken hold in Britain.

On the other side of the debate, designers joined in to commend Mercer on his paper and to share their frustrations with the limited British perspective of the industrial designer as a specialist. Remarking upon the low levels of publicity for the designer in Britain, designer Peter Hildesley remarked that he had worked in the US for two years, where 'nearly every manufacturer knew my work'; 'in this country no one has heard of me'.[49] Proponent of design reform, John Gloag, also in the audience, added:

> The buying public in America recognises that a stove or some other domestic appliance, designed by a good man, is a good thing. American manufacturers realise that by bringing the designer out of the anonymity in which he works here by using his name and putting over to the public as a specialist who can give them a better looking thing and a cheaper thing, they are using a selling asset, which, for the last century, we have neglected.[50]

The British fixation on the role of Consultant Designer was, above all, an awe-inspired reaction to the public visibility of the industrial designer.

Surprisingly, the role of the publicist, though critical to this visibility, was never mentioned in the discussion.

The Consultant Designer provided a useful prism through which the British could reflect upon their own problems. On closer reading, Mercer's perception of the situation of the consultant in the US was peppered with factual inaccuracies and misrepresentations. For instance, he incorrectly referenced Harold Van Doren ('Carl Van Doren'), while referring to the SID's success in establishing 'rigid training and examination as qualification conferred by other leading professional bodies', a much overblown assessment, probably informed by the publicity offices of the consultancies described. In addition, design reformers, including Mercer, placed a great stock on the education of the industrial designer in the US, apparently paying little attention to the fact that most of the 'big- name' Consultant Designers listed had no specific training in industrial design, many having a background in theatre and the graphic arts. Mercer also did not touch upon the consultant's dependence on the corporation and his isolation from questions of social responsibility – issues that would soon become subject to critique in the US and Britain.

Anyone listening to Mercer's speech would have been forgiven for thinking that design consultancy work did not exist in Britain at this time, but this was evidently not the case. Sir William Crawford, Richard Lonsdale-Hands, Hulme Chadwick, Kenneth Holmes, FHK Henrion, Willy de Mayo and Gaby Schreiber had all set up independent design consultancy offices in Britain before the war. While these practices were said to have resembled the 'American model' of consultancy (headed by an individual), the Design Research Unit (DRU), founded in 1942, constituted something distinct in its form, structure and organizational ethos. This practice had evolved from the studio model of Bassett Gray in 1931, to become the Industrial Design Partnership (IDP) in 1936, composed of Misha Black, Thomas Grey, Milner Gray, Jesse Collins, Walter Landor and James de Holden- Stone. Together, they focused on packaging, not as a 'by-product of advertising, but as a discrete professional practice', emphasising the relationship between specialized expertise and professionalism.[51] Announcing the formation of the IDP in *Advertiser's Weekly* in 1935, Thomas Gray said:

> This change will give the organization a central core capable of tackling design problems in the same way that an architectural partnership tackles building problems. In the past Bassett Gray was really an agency distributing the work of a group of free-lances. The members were not legally bound together as the six partners will be.[52]

Gray emphasized the value of specialization in the IDP model, insisting that 'no one designer can produce every sort of design', a statement in line with the British preference for representing the designer's role as part of a team, rather than on the basis of individual contribution. This aspect of the practice was further emphasised as it became the DRU in 1942, with its supposedly flat organizational structure, strict democratic lines, with 'no boss' and 'open' monthly meetings to which members of the press and other interested individuals were invited to participate in a critique of ongoing projects, presenting a strong image of design practice as a 'co-operative activity'.[53] In doing so, it presented a model of consultancy work that complemented the British preference for specialism over generalist expertise and the values of teamwork over individualism, assets that persisted well after the adoption of the title of General Consultant Designer in Britain.

The General Consultant Designers Group

After the war, hierarchical attitudes towards the staff and Consultant Designer that had been circulating in the US had gripped professional committees set up to reform the industrial design sector. A report on the training of the Industrial Designer prepared by the training committee of the Council of Industrial Design (CoID) in July 1946 remarked upon the significance of the 'profession of design consultant that has come into being in this country and the US'. The report focused on the 'urgent need for good designers', with 'good' here taking on the qualitative force of design reform. It drew upon a new categorisation between the designer as an 'expert professional' capable of consultancy to industry, to the 'designer-craftsman' and the 'rank and file of the design rooms in industry':

> It is however the relatively small number of designers of the first class whose ideas and influence will inform industry as a whole. The provision of better designers at the top end of the scale will rapidly affect the general standard and it is therefore with the training and activities available to those who are potentially capable of reaching that level that we are especially interested.[54]

The report presented the qualities of exclusivity, elitism and individuality as the route to professionalization of industrial design, as the influence of the US profession was undoubtedly now felt in the report. Importantly also, the report committee represented a cross-section of both design and manufacturing, with Milner Gray of the SIA, Margaret Allen of the CoID, ceramics manufacturer Josiah Wedgwood and textile manufacturer Thomas Barlow representing a shift in the relationship between the two factions, at least within the circles of design reform.

When the Consultant Designer role was formally adopted in Britain, it was under the somewhat awkward title of the 'General Consultant Designer', carrying into British industrial design discourse the distinction of working as a generalist for the first time. The SIA established the General Consultant Designers Group in 1953, to 'bring together for purposes of discussion, exhibitions of work, public relations generally and for the formulation of Codes of Conduct, those members of the Society who, having had specialized experience in one branch of design are in practice as both general and Consultant Designers'.[55] Membership was by application only, with stringent requirements that members had more than seven years of experience working across more than one design field. These fields were: graphic design, construction design, product design, product design engineering and 'miscellaneous skills'. Members were to be judged on their individual achievements and talents, rather than the output of the team they led, putting a renewed emphasis on the designer's individualized identity. They held their meetings at the Royal College of Art (RCA), followed by lunch at the Arts Club, and formed an elite identity within the SIA, often taking on the role of organizing events, including the Annual Dinner. All of its members were Fellows of the Society and only two were women (both Austrian émigrés, Gaby Schreiber and Jacqueline Groag). Indeed, a significant proportion were European émigrés, putting the Consultant Designer's claims to bringing an 'outsider's perspective' in a new light. As émigré FHK Henrion put it in the Design and Industries Association (DIA) Yearbook:

> Imagination cannot create anything which has not already existed; the creative act can only bring about new constellations of old ingredients or approach things from an unusual angle. It is obvious in this context that the Consultant Designer, since he comes from the outside to the problem and therefore approaches it with a fresh eye, is at times better placed to find an unorthodox solution than the people inside the organization, including specialist designers, who have grown familiar with their problems and orthodox solutions.[56]

This proclamation of special status did not sit comfortably with cultures of professionalism in British industrial design. Indeed, the exclusivity and narrowness of the Society of Industrial Artists (SIA) Group made it unpopular within membership. In his history of the Society of Industrial Artists and Designers (SIAD), designer and architect James Holland pointed out three criticisms of the group. First, as he put it, 'to some extent every designer is, or may be a consultant, and that often over a field wider than his own practice'. Secondly:

> to offer general consultancy services must be to encroach on all areas of specialist practice with the unavoidable risk of being treated as no more than an entrepreneur, and thirdly, a closed group of general consultants is apt to fix admission to its ranks in terms of its own image, since who else is to say who qualifies? There was little regret for the eventual disappearance of the Group, and there is no evidence that design practice in general, or the Group members themselves are any the worse or have had to restrict their range of practice.[57]

This critique reveals the cultures of professionalism in British design practice that were culturally opposed to the individualistic notion of the designer as a consultant and to the entrepreneurial ideas of design as a commercially oriented practice. As Holland suggests, the 'importation' of the title of consultant was something of a shallow exercise that worked to promote the status of those who worked under such a title. Indeed, the formation of the group within the SIA may have been no more than a publicity exercise. It is surely not coincidental that the Society established its first Public Relations Committee in the same year as the General Consultant Designers Group. Taste, experience and 'cultural knowledge' were values attributed to members of this group, underlining its significance as a marker of distinction, rather than a disciplinary or skill-based role.

The application of the title General Consultant Designer had the desired effect of attracting the attention of the media for the small minority of London-based practitioners working under the title. In February 1960, the *Observer* stated that 'the emergence of General Consultant Designers as a group in Britain within the past year is an indication that we are slowly coming round to the idea of design as a necessary element in daily life'.[58] Fashion and lifestyle magazines featured Consultant Designers including FHK Henrion and Gaby Schreiber, 'at home', presenting a highly idealized vision of the designer's role and identity. In 1966, Henrion confirmed this when he stated:

> Design has become fashionable. The leading Sunday papers devote 2–4 pages to it and even some of the serious weeklies feature design from time to time with critical appraisals of design problems. Hollywood produced a film where the hero was an industrial designer and even some public schools careermasters consider it as a possible profession for some of the boys leaving, not necessarily restricted for those who are otherwise hopeless and have merely a faculty for drawing.[59]

Here, Henrion reinforces the importance of self-image and 'fashionability'; features of professionalization in design that point towards the significance of the media as an agent. It also reveals the class-based social dynamics of professionalization in Britain, as Henrion proudly refers to the acceptance of the profession within public schools as an indicator of success. Referring directly to Henrion, the *Sunday Times* stated in 1960 that 'industrial design has developed into one of the glossiest professions and the most quasi-scientific of the applied arts and acquired a truly Hollywoodian glamour', referencing the visual allure of this new, glamorous profession as well as the American origins of its image.[60] The language used to define Henrion's work, by journalists at the time and historians thereafter, focused on the 'totality' and 'professionalism' of his work. Henrion, it has been said, was 'the complete designer'; a descriptor that mirrors the holistic qualities of the pioneering, 'industrialized' US design consultant.[61]

Reflecting the greater visibility that had been applied to the Consultant Designer in the US, a small minority of British designers were also now featured in fashion and lifestyle magazines in editorial and advertisements that drew on their self-image as taste-leaders to sell consumer products. In 1955, Robin and Lucienne Day, for instance, featured in an advertisement for Smirnoff vodka, in which, significantly, their status as designers of 'highly individualistic' furniture and textiles was referenced alongside their roles as glamorous hosts in 'entertaining visitors from Europe and America' (see Colour Plate 6). Consultant designer Gaby Schreiber, discussed further in the next chapter, also featured in an 'advertorial'-style article for the Jaguar Mark VII car, in which she is described as a 'very capable woman of affairs'. The article draws upon Schreiber's demanding work schedule, which involved long business trips and drives between London and her weekend home in Sussex, presenting a highly aspirational designer lifestyle for the *Vogue* reader.[62]

'The Anonymous Designer'

In 1957, the *Times* newspaper published a public letter from a 'Special Correspondent' entitled 'The Anonymous Designer', directly addressing the identity of the British designer in industry:

> Something like a formal recognition has been granted to the industrial designer in the past few years. His activities have lost the marginal character which made them seem not quite an art and not quite a profession. Many writers have championed his right to be regarded as an artist. But his admission to the kind of professional niche occupied by the architect has yet to be accomplished.[63]

This letter puts in perspective the limited achievements of the General Consultant Designer identity as a strategy through which to make the designer more visible in Britain, where they were still regarded as essentially anonymous, undervalued and 'semi-professional'. Written 'anonymously', it is interesting to contrast this piece with 'Both Fish and Fowl', the extravagant portrait of the US industrial design consultant in *Fortune* magazine (1934). Both were directed towards the client as the principal reader. Nevertheless, the bold and confident representation in *Fortune* puts into perspective the highly cautious, tentative and insecure nature of design discourse in Britain, which seemed to be stuck forever in a circular conversation about its position between architecture, engineering and advertising, between the 'new and old' professions.

Responses to 'The Anonymous Designer' letter were published from those working in architecture and engineering, as well as other designers. Alec B. Hunter, President of the SIA, responded to the letter with enthusiasm. He wrote to celebrate the 'diversity' of the industrial design profession in Britain and reflected on the eclecticism of the SIA membership, praising on the one hand 'the essentially personal work of the illustrator', alongside the 'designer director of an industrial undertaking concerned in the production of furniture, textiles, engineering products, printing and book production and so on'. Adopting the language of the US discourse, Hunter stated, 'Some members work as staff designers and others as independent consultants and both are, as your Special Correspondent says, regarded "as essential instruments in holding and extending markets"', showing sensitivity to both contingencies of the final paragraph of the SIA membership document. Nevertheless, in his final statement, Hunter hinted at the value of the generalist over the specialist, when he referred to the need for 'education which will not only provide mastery over technique but also be wide enough to include the "common factor" and a general background which will equip the designer to assume the authority and leadership which is now demanded'.[64] A further response to the letter was received from Richard Lonsdale-Hands, a designer said to have modelled his consultancy directly on the 'integrated' approach of advertising, design and marketing and who had long expressed admiration for American practice.[65] Adopting the language of the generalist, he wrote, 'We must be jacks of all trades …Only through working together can we become masters of our own.'[66] Lonsdale-Hands would come to represent the 'new-wave' model of consultancy in Europe that adopted an integrated model, with marketing, design and advertising.

Design integration

Ironically, by the time the Consultant Designer label had been adopted in Britain, the identity was reaching its limits in the US, as fewer corporations and businesses employed the designer as a consultant and moved towards in-house integration. In 1957, industrial designer Raymond Spilman conducted independent research into the integrated position of the Consultant Designer within the corporation. He listed eight major US corporations where the industrial designer was embedded within the corporate management structure. These were: the Rowe Manufacturing Co.; the Washburn Co.; the Container Corporation of America (CCA); Reynolds Metals; the Simmons Co.; Motorola, Inc.; the Underwood Corporation and Pickering + Co., Inc. This integration of the design consultant's role within the organization, shown here in the case of the Container Corporation of America, reoriented the question of agency inside and outside the corporation.

As Spilman identified, design was now an essential component in forming the 'public personality' of the company, a role that reoriented the object within design discourse. Industrialists, including Walter Paepcke at the

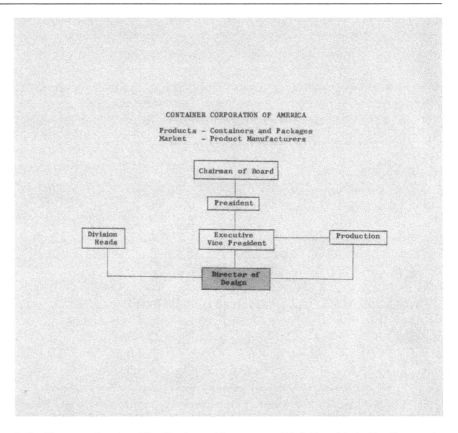

2.2 Diagram showing 'The Designer-Management Relationship in the Corporate Structure' within the Container Corporation of America (1957). Raymond Spilman Papers, Box 11, Folder 'The Designer-Management Relationship in the Corporate Structure-SAM Talk, 1957', Special Collections Research Center, Syracuse University Libraries

CCA, 'sought to establish an ideological rather than purely pragmatic relationship between design and business to imagine them as mutually constitutive rather than just expediently allied'.[67] As Egbert Jacobson of the same corporation put it, design was an essential component of the company, whether through the typography of the Annual Report or the design of the lockers and cafeterias. 'You cannot talk about design without suggesting integration. To integrate is to make whole. Design is only one important function of management but successful management itself involves the very highest type of integration.'[68] In 1959, editor Seymour Freedgood wrote in *Fortune* magazine, where the Consultant Designer's status had been so enthusiastically promoted in 1934, that 'probably no more than 300 firms in the US meet the generally accepted definition of an independent design office … The truth of the matter is that most popular ideas of the business derive from the spectacular and sometimes eccentric men who originated it, although these early practitioners have now faded away.'[69] At a meeting in June 1961, the IDI proffered, 'is the consultant designer going out of style?', a cutting but pertinent question that clearly took aim at the superior attitudes held within the consultant-dominated SID.[70]

The integration of design to the centre of business operations within the corporation presented opportunities and challenges for the formative members of the SID, for whom external consultant status had operated as their claim to exclusivity. Henry Dreyfuss warned against the integration of the designer into the corporation, stating 'there is a danger that the internal designer, no matter how good he is, may become frustrated by the office politics and dulled by the sameness of his work, so that in time he loses his eagerness and originality and takes the easy way. When this happens, he no longer creates, he is a captive.'[71] Historians have also gone along with this view. Writing in 1990, Richard Buchanan reflected on the absorption of designers into corporate structures as a loss of agency and independence, while Jeffrey Meikle has argued that design became 'smothered into a routine business function' after the Second World War.[72]

Many members of the SID pushed against this notion of 'captivity' inside the corporate design department, arguing that this view was bolstered by an overblown and egotistical idea of the designer's role. In an essay entitled 'Corporate vs Consultant' in *Industrial Design*, George Nelson, the designer and critic who had penned the article most associated with the dramatic entrance of the industrial design consultant to US business, critiqued the cultural superiority that was built up around the identity of the Consultant Designer, asking, 'has anyone ever heard … a designer refer to his client as a "captive" executive?'[73] This superiority, he argued, was also based on economic privilege, as hourly rates for consultants was estimated to be roughly three times that of the draughtsman-designer's.[74] Nelson argued that the Consultant's position of status was based on the perception of risk, whereby the 'risk-taking independent is superior to the individual who scuttles under the protecting wings of a benevolent corporation'.[75] The integrated model, he suggested, reshaped the identity of the Consultant Designer as an 'expert' to one of an 'attentive listen[er]', stating that 'with today's fuller integration of the skills available to companies, the indigestible "expert" has become something of a nuisance and a menace to boot', a statement that anticipated the declining significance of professional expertise more broadly.[76]

The systematic designer

The integration model in the US was paralleled in Britain by the publication in *Design* magazine of a series of articles, 'Systematic Method for Designers' (1963–1964). Developed by L. Bruce Archer, the theory aimed to do away with the 'old notion of design as a largely intuitive act', instead leading through a systematized, logical tool to guide the designer through the production team. Archer's writing was supplemented with a series of flow-chart diagrams representing design as a specialist ally of the industrial process. His thinking was phenomenally successful, inspiring an internationally influential design research movement that repositioned design as a scientific process, reviving and reinforcing the British preference for teamwork and specialization. Michael Farr's work on Design Management, also published first in *Design* magazine (1965), followed as an extension of this philosophy, defining the role of a manager in the design process, as a co-ordinator of the specialized work of an industrial design team for a client or corporation. Both ideas removed agency from the individual designer and distributed it more evenly across the team and within business.[77] Farr stated in *Design* that his model of design management was governed by the 'underlying premise: if designers are good at designing they should not have spare time to manage the ramifications of their design projects'. It aimed, he said, to harness the role of design as a 'unique factor in competition'.[78] In October 1965, Farr gave a report to the newly formed Industrial Designers Society of America (IDSA) on 'Design Management in British Industry'. Farr and Archer toured extensively together to promote their ideas about systematic design thinking, finding a receptive audience in the US, where ideas about integration of design and management within the expanded notion of design as a 'service' were circulating.[79]

While Archer would later claim that it was never his intention to eliminate or downgrade the qualitative judgement of the individual designer, some of his early writing in *Design* did have this effect. Writing in 1960 to review an exhibition of Consultant Designer FHK Henrion at the Institute of Contemporary Arts (ICA) gallery in London, Archer revealed his scepticism for the 'General Consultant Designer' identity:

> The inference of the exhibition appears to be that the general consultant designer is a special kind of designer who can turn his hand equally to designing a firm's letterheads, products, trademarks, exhibitions and packaging. Is this really possible? If one is to go by the evidence of the exhibition, the answer is 'no'.[80]

Archer's reading of the Consultant Designer's role focused on the tasks and skills involved, rather than the identity or personality that had been promoted through media in the US and now in Britain. For Archer, it was only in the context of co-ordination that the role made any meaningful sense: 'While specializing in one field of design, the [General Consultant Designer] is capable of guiding other designers in the project of a consistent image of a firm through its products, literature, advertising and show room design.'[81] His comments confirmed the continuance of British resistance to the 'special status' designation afforded through the title of General Consultant Designer, in many influential corners of the profession's media.

'New-wave' consultancies

In January 1964, Misha Black wrote to Anne Davies, secretary at the Industrial Design Institute (IDI), to ask to place one of his students in a consultancy in New York, stating, 'practically all of my students have one over-riding ambition. That is to gain some experience in the USA. The fact that so few do in fact come to your country is due only to their lack of funds and the few scholarships here which make that possible.'[82] Even if British design reform circles, organizations and publications (including *Design* magazine) warned against the adoption of American-style commercialism, these warnings had limited hold on an emergent generation of industrial designers. Indeed, by the 1960s, the 'American way' of doing business had become synonymous with economic reward, so that it was used as a shorthand way for individual designers to express commercial ambition. This was enhanced by the opening of the European Common Market in 1957, which marked an important turning point for a more open engagement in commercially competitive discourse in the British design profession (even though the country did not formally join the European Economic Community until 1973). This can be most clearly seen in the language and representation of two designers, James Pilditch and Richard Lonsdale-Hands, often presented within the media and in subsequent historical analysis as representative of a 'new wave' of commercially adventurous design consultants in Britain.

Founded in the late 1930s, the Lonsdale-Hands organization grew to have offices in 70 different countries with 500 employees. By the 1960s, the organization formed a 'pattern of companies' that dealt with research, marketing, design, advertising, merchandising and promotion departments working together in what Lonsdale-Hands describes as 'product strategy'.[83] For design historian Penny Sparke, Lonsdale-Hands's practice represented the first discrete design organization in Britain that was explicitly modelled on American management consultancy techniques. Indeed, his obituary in *Design* in 1969 makes this connection:

> Britain has lost her nearest equivalent to a first generation industrial designer, for Lonsdale-Hands was cast from the same mold that created the early American pioneers. Like his US counterparts of the late twenties and early thirties, he was as much a salesman as a designer, as much businessman as artist.[84]

Representations of Lonsdale-Hands in the media frequently focused on this perception of cultural distance between US and British business practices, and this was a position Lonsdale-Hands himself exploited, cultivating an image of British gentility for the US media on the one hand

and buccaneering entrepreneurialism in Britain, an inversion of the roles adopted by Raymond Loewy. He was not the first designer to manipulate his media persona in this way. David Ogilvy, 'the most celebrated advertising man of the 1960s, the British ad man who had it big in the States',[85] successfully 'played on his Home Counties heritage and Oxford education to great effect'.[86] Lonsdale-Hands cultivated and used his marketing skills to attract the attention of the US and British trade press, adopting different personas for each. In 1961, an article in *Advertising Age* presented a glowing feature on Lonsdale-Hands in advance of the opening of his art exhibition at the Hirschl & Adler gallery in New York:

> In the past ten years he has put together the Lonsdale-Hands organization, a group of interrelated companies covering all phases of marketing, advertising, PR, market research, package design and international marketing – as well as interior decoration. Greenly's is the agency wing. The Lonsdale-Hands organization reported billions of $16,272,000 [sic] for 1960.[87]

The same year, Lonsdale-Hands penned an article for the *American Artist* magazine, in which he announced that 'Industrial design in the British Isles has come of age.'[88] A promotional video for his organization, sketching 'the activities of an unusual marketing organization', also filmed that year, shows Lonsdale-Hands performing the image of the English gentleman-designer. The video opens in London, as the narrator comments, 'where its British headquarters are located. It respects some of the old traditions but looks to new skylines.' The film moves between the research, marketing, design, merchandising, advertising and promotion departments within the organization, which Lonsdale-Hands explains are 'dovetailed' and 'co-ordinated' activities. However, on closer inspection, Lonsdale-Hands was more than an advocate of the integrated model of American design consultancy. He was an internationalist, whose approach anticipated the merging of markets at a time when opening of the European Common Market was a big discussion point in the British and US design industry. In a particularly revealing interview in the *Christian Science Monitor* in 1961, Lonsdale-Hands was critical of the US approach, which he argued was over-reliant on research. Instead, he said:

> I think the West is slightly running out of ideas. That is one of the reasons why I'm looking forward so much to stopping off in Japan. There is nothing that can come from Red China, but from free Asia we may get much if we open our eyes to receive it.[89]

This bold statement puts in perspective the representation of Lonsdale-Hands 'cast in the mold of the early American pioneers'. His approach could be more accurately characterized in the context of free-market libertarianism. He was similarly critical of British business practices. As an article in *Time & Tide*, a British literary review magazine, in April 1962 stated, 'Mr Lonsdale Hands would like to see words like "exports" and "overseas" and "foreign" banned. He feels they conjure up the wrong images and allow the British to continue to believe "Fog in channel: Continent isolated." There are no export markets. There are simply needs awaiting satisfaction.'[90] In many ways, his approach anticipated the neoliberal turn in design culture.[91]

The formation of the European Common Market forms an essential context against which to understand the success of the 'new-model' consultancies that included the Lonsdale-Hands Organization and Allied International Designers, founded by James Pilditch. As Lonsdale-Hands put it, 'Exports are going to be terribly important, vital to Britain. We are going to be pitched into the Common Market by circumstances as well as politics. We must be ready for it.' Referring to his recent opening of an office in Geneva, he stated, 'I am in Switzerland now for that reason. This Plan International I see as a springboard for British businessmen into the Continent.'[92] Allied International, founded by James Pilditch, was the first quoted design consultancy on the stock market, which 'set the pace for other consultancies' that emerged in Britain.[93] Originally trained as an art historian, Pilditch had moved to New York and Toronto in the 1950s, where he had learnt about the structure and organization of consultancies there. He was passionate about the 'Anglo-American partnership', which he claimed to be 'as natural as breathing'.[94] He returned to London with a dynamic view of how design should be structured economically. His books *The Silent Salesman* (1961), *Talk About Design* (1976) and later, *I'll be over in the morning* (1990) represented a business-centred perspective on the role of design, particularly connected to an international market economy.

While a passionate advocate of transatlantic culture, Pilditch frequently referred to the new direction in British consultancies in a European perspective. As he wrote in *The Silent Salesman*: 'It is becoming a more total service, more powerful and authoritative ... In Europe as in the United States, industrial designers are becoming what *Fortune* called "the trained dreamers" of industry. They are no longer a 'glorified art studio round the corner', a cutting reference to the DRU model described earlier in this chapter.[95] In 1965, he co-wrote a design manual, *The Business of Product Design*, with Douglas Scott, an industrial designer who set up Raymond Loewy's London practice between the wars. In the book, Scott and Pilditch pay homage to the American origins of design consultancy, while noting the distinctive structure and outlook of the British model, which they contextualize in a European context. 'After a sluggish start, British designers are now being used more and more to create goods for mass sale ... It may be an American invention, but in Europe the design organization, as such, finds purest expression in London.'[96] Pilditch identified the development of a new kind of design organization in Britain that had moved beyond the model of the American consultant-led operation. 'What distinguishes the modern design organization? Recognition that one man cannot be expert in everything.'[97]

US industrial designers also situated the British industrial designer within a European context. As Henry Dreyfuss put it in 1946:

> The industrial designer working in the United States should not be compared to a European designer, who is so often a craftsman making only a few kind of precious products, usually for a sophisticated audience. Here in this country, we design products to be produced by the hundreds of thousands, often millions.[98]

In spite of this dismissal, by 1957, his company published a press release explaining their decision to open a London office, stating that the London and New York offices would serve 'complementary and reciprocal functions'. While the New York office would be a 'fact-finding office for British industry wishing to expand sales in American markets', the London office would focus on engineering and development projects.[99] Donald Deskey also pointed to practical benefits of manufacturing in Britain, where the delivery time for packing was much faster. His opening of an office, alongside other design consultancies at this time, points to the increased contact between the two markets.[100] An article in the American Society of Interior Designers (ASI) Newsletter (August 1964) announced the speakers for a forthcoming conference, 'Science, New Parameters for Industrial Design', in which Misha Black headed a list of impressive speakers: 'his organization has impressed many visiting American Industrial Designers (notably the late Walter Dorwin Teague) with the scope of its enterprise and the outstanding quality of its work'.[101] The British Trade Center was established in New York in 1967, 'an organization to help promote British goods in America', aiming to bring British exporters, particularly medium-sized firms, into direct contact with American markets.[102] In 1968, the Conran Design Group landed 'possibly the largest ever contract between a US firm and British design team' and Macy's ran a 'Best of Britain' fortnight in its stores.[103] This economic exchange can also be contextualized in relation to the parallel 'American invasion' of British advertising as part of the international expansion of world trade.[104] Anecdotal evidence within the archives of designers and their companies reveals the 'invasion' of similar business practices.[105] Convergence between Britain and the US was indicative of much wider patterns of globalization and the growing internationalism of design, rather than the joining of professional ideals between the two countries. Indeed, as the previous chapter discussed, British and American industrial designers were engaged in a process of cultural exchange, informed by flattery and critique that moved back and forth in both directions. As Chapter 4 will further argue, the two cultures continued to thrive through a perception of cultural distance, divergence and distinction.

Conclusion

If, as F. A. Mercer said, professions need pioneers, then the profession of industrial design found a highly visible and alluring image in the identity of the Consultant Designer, a self-appointed title that not only enabled the industrial designer to gain greater visibility and professional status, but in turn placed the industrial designer in an intermediary position, to sell an idealized professional self-image to the public. These designers – who reached 'household-name' status in some cases, while simultaneously forming a significant minority within the profession – secured a space in the history of the profession that is disproportionate to their influence. Indeed, the fact of their prolific self-promotion makes the Consultant Designer one of the profession's most unreliable narrators, as is notable in the heavily mythologized accounts presented in their autobiographical works and manuals. Professionalism served a very particular function to this elite group, serving an instrumental function to build an impressive self-image in the eyes of the client they depended upon for their income. Accordingly, institutions including the SID and, to some extent, the SIA, set up and adapted their professional codes and ideals to fit their own requirements. Nevertheless, as Chapter 4 will explore further, this exclusivity faded over time, and became discredited through social critique that attacked the premise of privilege and professional expertise the Consultant Designer had so determinedly and visibly sought to project.

In the identity of the Consultant Designer, American and British design share a history, but the details and context of this history have been misrepresented. According to the dominant narrative in British design history, the General Consultant Designer was the successful result of years of persuading by propagandists, including John Gloag, Mercer and Raymond Loewy, 'to inform a British audience about what was happening on the other side of the Atlantic and to encourage Britain to participate in the struggle to establish the industrial designer as a valid and useful professional'.[106] Under this analysis, it seems that consultancies in Britain only find their momentum once the American approach has been accepted. However, American culture, commercialism and business practices had 'become the norm against which many Europeans came to judge their own way of life'.[107] As this chapter has argued, the Consultant Designer was significant for a British audience first and foremost in a representative capacity, standing for commercialism and greater visibility, qualities that were felt to be lacking at home. It also signified a sympathy between manufacturers and designers that did not define the relationship in Britain. The US 'pioneers' functioned as a visible personalities through which to convey and express alternative norms, identities and business approaches.

In the context of the internationalization of business markets, British design practice was finding its own way to accept these values, a trend that continued apace when Britain finally entered the Common Market in 1973. The Consultant Designer title, in this context, represents a relatively shallow attempt from within the SIA to accelerate this process. The formation of the General Consultant Designers Group was a public relations exercise for the Society, which grouped together independent, freelance designers and group-based practices under a working title for the first time. The more significant change occurred after that when consultants including FHK Henrion and HA Rothholz, as well as group practices including the DRU, moved from 'services founded on the deep-rooted knowledge of a single design craft' to a 'more market coordinated service that focused on the concerns of a company's collective design policy'.[108] Ultimately, the 'pioneering' model of the individual master-of-all-specialisms failed to convince a profession that had been brought up to believe in the value of specialized expertise and teamwork as core values.

Walter Dorwin Teague's death in 1960 tolled the 'end of an era' for the independent design consultant. The merger of the IDI and SID in 1965 to form the IDSA represents the closure of this so-called 'golden age' for the professional identity of the independent design consultant, although its image holds strong. In spite of this, it is still the 'big-name' Consultant Designers who have found a privileged space within the history of the profession. The pioneering ideals and bold expression of professional self-determination secured a space in the history of design – and the cultural imagination – that endures in the present day. Moving beyond this, the next chapter looks beneath the mythologized privilege of the male design consultant to look at 'other spaces' of the profession, including the mechanism of publicity in the professional identity of the 'woman designer' and organizational and administrative roles within the profession.

Notes

1 'Then an unorganized following and only third, the regulation of its activities', F. A. Mercer, Editor of Art and Industry, Sixteenth Ordinary Meeting, Wednesday 11 April 1945, printed in the *Journal of the Royal Society of Arts* (8 June 1945), p. 349.
2 Penny Sparke, 'Introduction', in *Consultant Design: The History and Practice of the Designer in Industry* (London: Pembridge Press, 1983), p. 3.
3 See David Wang and Ali O'Ilhan, 'Holding Creativity Together: A Sociological Theory of the Design Professions', *Design Issues*, 25:1 (2009), 15.
4 See Paddy Maguire and Jonathan Woodham (eds), *Design and Cultural Politics in Post-War Britain: The Britain Can Make It Exhibition of 1946* (London: Leicester University Press), pp. 34–5. Guy Julier, *The Culture of Design* (London: Sage, 2013), p. 32.
5 Christopher D. McKenna, 'The World's Newest Profession: Management Consulting in the Twentieth Century', *Enterprise & Society*, 2:4 (December 2001), 673.
6 Michael Ferguson, *The Rise of Management Consulting in Britain* (London: Routledge, 2002), p. 2.
7 Ibid., p. 13.
8 Ibid., p. 43.

9 Samuel Haber, *The Quest for Honor and Authority in the American Professions, 1750–1900* (Chicago: University of Chicago Press, 1991).
10 Craig Robertson, '"Will Miss File Misfile?", The Filing Cabinet, Automatic Memory and Gender', in Sarah Sharma and Rianka Singh (eds), *Re-understanding Media: Feminist Extensions of Marshall McLuhan* (Durham, NC: Duke University Press), pp. 152–3.
11 Jan Logemann, *Engineered to Sell: European Émigrés and the Making of Consumer Capitalism* (Chicago: University of Chicago Press), p. 10.
12 Ferguson, *Rise of Management Consulting*, p. 2.
13 George Nelson, 'Both Fish and Fowl', *Fortune* (February 1934), 40–4, 88–94.
14 Alice Twemlow, *Sifting the Trash: A History of Design Criticism* (Cambridge, MA: MIT Press, 2017), p. 55.
15 Ivan Illich, *Disabling Professions* (London: Marion Boyars, 1977), pp. 11–12.
16 Nelson, 'Both Fish and Fowl', 88.
17 Ibid.
18 Ibid., p. 43.
19 Ibid., p. 40.
20 Egmont Arens, File 31, Job Descriptions, c.1945. Arens papers, SCRC.
21 Loewy office Job Specifications, c.1945, Arens papers, SCRC.
22 Brook Stevens, Job Specifications, c.1945, Arens papers, SCRC.
23 Russel Wright, 'Industrial Design for Today's Living', c.1950s., Box 38, Folder 2, Wright papers, SCRC.
24 Ibid.
25 Ralph Caplan, Betsy Darrach and Ursula McHugh, 'Publicity for a Profession', *ID Magazine* (October 1960), 74–85. Raymond Loewy Archive, CHMA.
26 Ibid.
27 In a letter to Don McFarland, Teague stated: 'No consultant designer can afford to hide his light under a bushel.' Walter Dorwin Teague to Don McFarland (4 May 1960), IDSA papers, Box 62, Folder 9, SCRC.
28 Richard Marsh Bennett, 'The Education of the Industrial Designer', in Richard Marsh Bennett and Walter Dorwin Teague, *Good Design is Your Business: A Guide to Well Designed House-hold Objects Made in U.S.A.* (Buffalo, NY: Buffalo Fine Arts Academy, 1947), p. 18.
29 J. Gordon Lippincott, *Design for Business* (Chicago: Paul Theobald, 1947).
30 Arthur J. Pulos, *Opportunities in Industrial Design Careers* (n.p.: Vocational Guidance Manuals, 1970), p. 155.
31 Russel Wright, 'Industrial Design for Today's Living', Wright Art School Lectures, Russel Wright 38_2, SCRC.
32 Raymond Loewy, *Never Leave Well Enough Alone* (New York: Simon & Schuster, 1953). Walter Dorwin Teague, *Design This Day: The Technique of Order in the Machine Age* (New York: Harcourt, Brace, 1940). These references were also made in the 'Fashions of the Future', US *Vogue* (February 1939), described in the previous chapter.
33 See Chapter 3 on Betty Reese, 106–111.
34 Arthur BecVar, General Electric (20 June 1962), Box 5, 'Talks 1961–1962', Raymond Spilman papers, SCRC.
35 Pulos, *Opportunities*, pp. 59–60.
36 The SID pursued a close relationship with the AMA from its institution and paid for this research. The AMA was founded in 1923.
37 American Business and Industrial Design, American Management Association and Society of Industrial Designers (1947), Arthur J. Pulos Archive.
38 Eastman Kodak report (April 1954), Raymond Spilman papers, SCRC.
39 Bourdieu first used the term to refer to a new 'class' of 'petit bourgeoisie working in occupations involving presentation and representation, providing symbolic

goods and services'. Pierre Bourdieu, *A Social Critique of the Judgement of Taste* (Cambridge, MA: Harvard University Press, 1984 [1979]), p. 359.
40 Keith Negus, 'The Work of Cultural Intermediaries and the Enduring Distance Between Production and Consumption', *Cultural Studies*, 16:4 (2002), 505–15.
41 Roland Marchand, *Creating the Corporate Soul: The Rise of Public Relations and Corporate Imagery in American Big Business* (Berkeley: University of California Press, 2001), p. 276.
42 Brochure for the Society of Industrial Artists (December 1945), CSDA.
43 Marsh Bennett, *Good Design is Your Business*, p. 21.
44 John Gloag, *The Missing Technician in Industrial Production* (London: Routledge, 1944), p. 9.
45 Mercer, Editor of Art and Industry, p. 343.
46 Ibid.
47 Second only to his American citizenship. Raymond Loewy, Transcript, 'I'm an American', US Department of Justice, Washington, DC, Sunday 11 May 1941, National Broadcasting Company, New York, Station, WJZ, Raymond Loewy, press clippings, HMA.
48 Mercer, Editor of Art and Industry, p. 343.
49 Peter Hildesley, quoted in Mercer, Editor of Art and Industry, Sixteenth Ordinary Meeting, Wednesday 11 April 1945, printed in the *Journal of the Royal Society of Arts*, 8 June 1945, p. X.
50 John Gloag, quoted in ibid., p. X.
51 S. L. Righyni, *Boxmakers Journal and Packaging Review* (1931). Misha Black Archive, AAD/1980/3/6.
52 'Bassett Gray Becomes a Partnership, New Firm Replaces 14 Year Old Organization' (23 May 1935). Walter Landor Archive, Landor Design Collection, NMAH.
53 Avril Blake and John Blake, *DRU: The Practical Idealists* (London: Lund Humphries, 1964), p. 26.
54 Report on the Training of the Industrial Designer (July 1946), Design Council Archive, UBDA, p. 13.
55 *SIA Journal* (October 1953), Box 95, CSDA.
56 DIA Yearbook (1959), DIA Archive, AAD/1997/7/120, VAAD.
57 James Holland, *Minerva at Fifty: The Jubilee History of the Society of Industrial Artists and Designers 1930 to 1980* (Westerham, Kent: Hurtwood, 1980), pp. 6–7.
58 Patience Gray, *Observer* (7 February 1960), FHK Henrion, press clippings, UBDA.
59 'The Designer as a Problem Solver' (1966), FHK Henrion, speeches, UBDA.
60 Robert Harling, 'Stylists of Industry', *Sunday Times* (1960), FHK Henrion Archive, UBDA.
61 Adrian Shaughnessy, *FHK Henrion: The Complete Designer* (London: Unit Editions, 2013).
62 Margaret Jennings, 'The Woman at the Wheel', *Vogue* (June 1956); Gaby Schreiber scrapbook, VAAD.
63 A Special Correspondent, 'The Anonymous Designer, Seeking a New Status in Industry', *Times* (9 January 1957).
64 Alec B. Hunter, 'The Anonymous Designer', *Times* (14 January 1957), p. 9.
65 Reporting on a recent trip to the US in the *SIA Journal*, Richard Lonsdale-Hands noted the 'very prosperous industrial designers' he found there, *SIA Journal* (May 1946).
66 Richard Lonsdale-Hands, letter, 'The Anonymous Designer', *The Times* (16 January 1957), p. 9.
67 Gordon-Fogelson, 'Vertical and Visual Integration'.
68 'Design as a function of management', American Ceramic Society Design Division *Newsletter*, 1 (April 1956), IDSA Archive, SCRC.
69 Seymour Freedgood, *Fortune* (1959), quoted in Sparke, *Consultant Design*, p. 37.

70 Allegeny Chapter meeting (9 June 1961), talks, 1961–1962, Raymond Spilman papers, SCRC.
71 Henry Dreyfuss, 'Industrial Design: profile of an organization', typescript of a talk given at Harvard University Graduate School of Business Administration, Cambridge, Massachusetts' (n.d.), Henry Dreyfuss Archive, Series 4.1, Box 1, CHMA, p. 12.
72 Richard Buchanan, 'Toward a New Order in the Decade of Design', *Design Issues*, 6:2 (Spring 1990), p. 80. Buchanan quotes Meikle in this article, p. 71.
73 George Nelson, *Problems of Design* (New York: Whitney Publications, 1960), p. 29.
74 Ibid., p. 30.
75 Ibid.
76 Ibid.
77 Archer was explicit that the aim of the Systematic Design Methods movement was 'not about helping practising designers – I am doing this to completely understand the design process', S. Boyd Davis and S. Gristwood, 'The Structure of Design Processes: Ideal and Reality in Bruce Archer's 1968 Doctoral Thesis', in P. Lloyd and E. Bohemia (eds), *Design Research Society Annual Conference Proceedings*, 16, 27–30 June 2016, Brighton, UK. https://dl.designresearchsociety.org/drs-conference-papers/drs2016/researchpapers/95, accessed 2 May 2024.
78 Michael Farr, *Design Management* (London: Hodder & Stoughton, 1966).
79 IDSA Presidents Report (25 October 1965), 'Michael Farr, London, England, gave a report on Design Management in British industry', Folder 10, IDSA Archive, SCRC.
80 L. Bruce Archer, *Design* (1960), quoted in David Preston, 'The Logic of Corporate Communication Design: Design Coordination and the Shifting Materiality of Practice for Consultant Graphic Designers in Post-War Britain, 1945–1970' (PhD thesis, University of the Arts London, 2018), p. 181.
81 Ibid.
82 Letter, Misha black to Anne Davies, IDI (13 January 1964), Leon Gordon Miller papers, SCRC.
83 1961 Promotional Video for the Lonsdale-Hands Organization, https://vimeo.com/28355748 (accessed 15 April 2024).
84 Unknown author, 'Obituary: Richard Lonsdale-Hands', *Design* (1969), in press clippings folder, Richard Lonsdale-Hands private archive, HAG.
85 Sean Nixon, *Hard Sell: Advertising, Affluence and Transatlantic Relations, 1951–69* (Manchester: Manchester University Press, 2016), p. 29.
86 Joe Bullmore, 'David Ogilvy: 'Life lessons from the godfather of advertising', *Gentleman's Journal*, www.thegentlemansjournal.com/article/david-ogilvy-life-lessons-godfather-advertising (14 April 2023), accessed 2 May 2024.
87 'Versatile Adman … Richard Lonsdale-Hands', *Advertising Age* (8 May 1961), Press Pack, HAG. In May 1961 the Hirsch & Adler Galleries, New York, staged a 'one-man-show' of sixty of Lonsdale-Hands's paintings.
88 Lonsdale-Hands, *American Artist* (November, 1961), p. 61, Box 43, Arthur J. Pulos Archive, AAA.
89 Londsdale-Hands, *Christian Science Monitor*, n.d. (c.1961), press clippings, HAG.
90 Unknown author, 'Big Business and the Common Market', *Time and Tide* (April 1962), HAG.
91 See Julier, *Culture of Design* and *Economies of Design* (London, Sage, 2017).
92 John Allen May, 'The Executive Outlook', *Christian Science Monitor*, n.d. (c.1961), Richard Lonsdale-Hands, press clippings, HAG.
93 Julier, *Culture of Design*, p. 32.
94 James Pilditch, *'I'll be over in the morning': A Practical Guide to Winning Business in Other Countries* (London: Mercury Books, 1990), p. 77.

95 James Pilditch, *The Silent Salesman: How to Develop Packaging that Sells* (London: Business Books, 1973), p. 146.
96 James Pilditch and Douglas Scott, *The Business of Product Design* (London: Business Publications, 1965).
97 Ibid., p. 76.
98 Response from Henry Dreyfuss to *New York Times* letters page (18 November 1962), Arthur J. Pulos Archive, AAA.
99 Dreyfuss, London office.
100 Donald Deskey Associates, press release, 'Restyling English Products for American Market' (8 November 1957), CHMRB.
101 ASID/IDI Newsletter (August 1964), National Meeting, Philadelphia, 15–17 October 1964, IDSA Archive, SCRC.
102 Arthur J. Pulos research notes, Box 43, Arthur J. Pulos Archive, AAA.
103 In connection with a 'best of British' fortnight' at the store. 'British design group invades US', press clipping, unknown source, Box 43, Arthur J. Pulos Archive, AAA.
104 Nixon, *Hard Sell*, p. 30.
105 For instance, in August 1971, a member of Walter Landor's consultancy travelled to London to meet F. J. Luck, the commercial director of Richard Lonsdale-Hands Associates, to discuss potential 'collaboration', meaning that they were looking for companies to buy. Luck explained that 'an exchange of directors between ourselves and them' would be the preferred method to pursue, reflecting the dynamic exchange between British and US creative work. In 1974, Paul Reilly, Director of the UK Design Council, wrote to Landor to ask if he was thinking of opening a London office, encouraging him to do so to stimulate the trend towards this kind of work. Landor Design Collection, Series 9, International Files, National Museum of American History, Smithsonian Institution, Washington, DC.
106 Sparke, *Consultant Design*, pp. 55–6.
107 Nixon, *Hard Sell*, p. 3.
108 Preston, 'The Logic of Corporate Communication Design'.

3
Women's work

Speaking at the opening of an industrial design exhibit at the new Illinois Institute of Technology in 1954, publisher of US *Industrial Design* magazine Charles Whitney said, 'Designers are about the most influential thing on the American scene – that is of course, aside from women and politicians.'[1] This 'othering' of women as a social category separate and distinct from 'designer' was commonplace and reflective of the culture of hegemonic masculinity encoded in the profession in Britain and the US. Early career-guidance literature pointed to the incompatibility of women's work within the 'hard-boiled' industrialized context of the factory or the design consultancy.[2] And yet – as these authors also had to recognize – women were 'ideally suited' to certain roles in industrial design and were key to its successes, even when not always visible.[3] Indeed, as Alice Twemlow has pointed out, Whitney's hard, industrialized view of the design profession as essentially male was complicated by the fact that *Industrial Design* magazine was edited by two women, Jane Thompson and Deborah Allen, who 'created a space for a distinctively American mass-market product design criticism, fuelled by their personal beliefs, intellectual backgrounds and experiences as both professional working women and home makers'. Nevertheless, the magazine reflected the masculine image of the profession, and 'not a single woman designer was profiled in at least the first decade of the magazine'.[4] This industrialized male aesthetic was visible in many of the cover designs for the magazine, including that shown in Colour Plate 7.

Given the tight relationship between professionalism and masculinity, it is no surprise that British design critic Reyner Banham dubbed it 'the most professional of industrial design magazines'.[5]

This chapter wrestles with this essential contradiction in the gendered identity of the industrial designer, looking more closely at the dynamics of a profession that projected a hyper-masculine self-image, but was dependent upon interaction with cultures of femininity in multiple ways. The first section looks at the identity of the 'woman designer', a media invention that conspired to reconcile for its audience the previously incompatible spheres of femininity and professionalism. Focusing on the representations of Gaby Schreiber, Freda Diamond and Florence Knoll Bassett, the chapter shows that in these specific cases, these women played the category of 'woman designer' to their professional advantage, as it placed them in closer contact with the female consumer and secured their authority as 'taste-makers', a signature feature of the Consultant Designer's identity, as discussed in the previous chapter.

Looking beyond the designer and into 'other spaces' of the profession, the remaining sections of the chapter focus on work within the profession predominantly performed by women. This includes publicity, administration and organizational roles; skills often performed by women within consultancies, offices and committees of professional and promotional design organizations. Radical work by feminist art and design historians since the 1960s brought to attention the historiographical absences of women who played important roles within the formative years of the profession, but owing to rather mundane reasons of archival absence, these stories still remain frustratingly underrepresented.[6] Writing in 2007, Lesley Whitworth and Elizabeth Darling articulated a 'growing engagement with the unknowable woman: the factory worker, the shop clerk, the housewife' in relation to design and the built space.[7] More recently, historians of work, including Andrea Komlosy, identified a further obstacle in the study of work from a gendered perspective, arguing that the very definition of what constitutes work in professional and non-professionalized fields was designed to fit the values of industrialized economies dominated by hegemonic masculinity.[8] This chapter responds to both of these works and attempts to redress that imbalance by examining 'intermediary' and 'administrative' roles within the design profession, finding that these were jobs mostly carried out by women, relegated as non-professional, or something 'other' to the professional working identities of the men they worked alongside.

Some of the evidence presented in this chapter has been pieced together from oral histories, uncatalogued archives and personal and private collections, a process that reflects the 'othering' of women's work within the historicization of the profession. This 'unknown woman' forms a dramatic contrast with the three female design consultants examined in the first and following section, whose place within the history of design was more confidently assured through their investment and co-operation with media and publicity during their lifetimes and subsequently, through archival institutionalization.

The 'woman designer'

The hyper-masculine identity of the industrial designer was further reinforced through the use of term 'woman designer', a title assigned to a small minority of women who achieved visibility and status within the profession in both Britain and the US in the post-war period. The use of this category by journalists and publicists at the time reflects the divisive effects of professionalization in design, through which, as Chapter 1 of this book explored, practices including crafts, textiles and ceramics were defined as non-professional and amateur in opposition to the professional, industrialized status of the industrial designer. The category was carried into the historiography of the profession, importing the 'problematic' identity of the 'woman designer'. When asked directly, many women expressed discomfort with this term, including Gaby Schreiber and June Fraser, design consultants working in Britain, who expressed frustration when asked to reflect on their experience as 'woman designers' working in a male-dominated field.[9] In some cases, design historians have further perpetuated the notion that women functioned as designers in an entirely separate sphere to their male contemporaries, 'hidden' and 'denied significance'.[10]

Closer scrutiny of the mediated identity of female design consultants practising in Britain and the US reveals some alternative perspectives on these problematic assumptions. These women, like their male contemporaries, utilised the media as a tool in professionalization, by employing publicists and PR consultants to promote their work. The identity of the 'woman designer' was invented through media platforms including fashion and lifestyle magazines, in parallel with the emergence of the 'New Woman', an aspirational category championed in the pages of such publications. As *Who's Who* noted in March 1956, women were featured in its US annual four times faster than the average for all its twenty-eight editions since its founding in the nineteenth century; the 'numerous opportunities for women to become prominent enough to become inquired about generally have resulted from the tremendous development of the applied arts, which has occurred in this country and the accompanying increase in opportunities for women in business, government and sciences'.[11] The collision of these categories presented a fitting framework through which to invent a glamorous, sophisticated and elegant image of the designer in the eyes of aspirant female readers.

American historian of technology Caroll Pursell has written of how women in the engineering profession after the Second World War had to strike a 'balance between femininity and professionalism'.[12] Media profiles and editorial features on the Austrian design consultant Gaby Schreiber, who emigrated to Britain during the inter-war period and established a highly successful career as a freelance consultant, and later her own consultancy Convel Ltd, presents a fascinating case study through which to view the these two seemingly disparate ideals in dialogue. Schreiber's carefully preserved press clippings, held in her archive at the V&A Archive of Art and Design, show the care she took to preserve and document this heavily photographed and mediated period of her life. Photographic portraits taken of Schreiber 'at work' by *Vogue* and *Tatler* photographers present her at her desk, poised and professional in appearance.

The photograph shown in Figure 3.1, collected by the Council of Industrial Design (CoID) as part of its Photographic Library, puts on display the representative capacities of photography that were employed in similar ways for Schreiber's male contemporaries to deliver a professionalized self-image. This official photograph, kept on file by the CoID for the purposes of representing her work to clients and also circulated to the media, confirms journalist Liz McQuiston's description of how Schreiber 'projects qualities of sophistication and professionalism in her physical appearance'.[13] Schreiber made quite specific demands upon the CoID with regard to how they photographed her design work, requesting a particular photographer (Sydney Newberry, freelance photographer for the *Architect's Journal*), so it is likely that she played an active role in the photograph's composition and aesthetic arrangement.[14] 'Perching' on the edge of her elegant desk chair, the image perfectly captures the 'balancing act' of professionalism and femininity implied by the title 'woman designer', while the averted gaze, 'Hollywoodian' in style, is markedly distinct from the direct gaze of her male contemporaries within the CoID's portrait collection (including James Gardner; see Figure 1.3). Here, feminine qualities of glamour, sophistication, elegance and beauty were mobilized in the construction of Schreiber's professional identity to captivating effect.

The consistency with which Schreiber projected this image, through interviews and through photographic portraits, suggests it is something she worked hard to maintain. During the period 1950–1960, she was photographed for fashion magazines including *Vogue*, *Tatler* and *Harper's Bazaar*. As *Industrial Design* magazine staff writers later explained, 'Women's page stories, which are thought by some PR people (generally men) to be pointless, are thought by others (generally women) to be darkhorse publicity', and Schreiber utilized this space with great success.[15] In 1968, *Harper's Bazaar* reflected on Schreiber's 'remarkable' ability to balance professionalism and femininity:

> Gaby Schreiber is two people. One of them is a woman, no longer very young, but still attractive and entirely feminine; men would rather think of her as a lover than a mother. The other is a designer, an experienced professional, successfully competing in a tough, competitive world. Somehow the two people get along extraordinarily well together. Neither one of them gets pushed into the background because the woman is a designer and the designer is a woman.[16]

Schreiber's physical appearance, which conformed to post-war aesthetic values of feminine beauty, glamour and sophistication, further enhanced her suitability for this idealized and aspirational representation. As the General Electric research report described in Chapter 2 argued, 'personal attractiveness to others' was considered a necessary qualification of

3.1 Gaby Schreiber, Consultant Designer, 1919–1991 (n.d., *c.*1950). Photographer: Gee & Wilson. Design Council Archive, University of Brighton Design Archives. Original image reference: GB-1837-DES-DCA-30-1-POR-S-16.

the Consultant Designer.[17] In 1951, the *Evening News* commented upon Schreiber's ability to 'combine femininity and efficiency':

> This slim, elegant woman of 35 with red gold hair and glowing dark eyes believes she is the only industrial design consultant in London. While we talked, Ninotchka her poodle dozed off in front of a log fire in Mrs Schreiber's office, with its white walls, blue curtains and claret-coloured chairs.[18]

Like those of her male contemporaries, the interior spaces of Schreiber's home and, in particular, her home office were rendered complementary to her identity as a professional designer, conveying to the reader expertise in matters of style and taste.

Schreiber was married three times and her marital status provided another important lens through which her professional identity was secured. An editorial feature on Schreiber in *House and Garden* (1958) presented the 'Town and Country' houses of Schreiber – here referred to by her marital name Mrs Fishbein – and her husband William Fishbein, also a designer:

> It's an inspiration to see actually achieved, a way of life that is so often an unrealized pipe-dream – a town and country life combining the best of both worlds. But here's a couple who have achieved it, thanks to the same careful planning that also makes their professional lives so successful. Mrs Fishbein, as Gaby Schreiber, one of the most prominent pioneers in the design of plastics, now designs the Bartrev Furniture, for which her husband William Fishbein both designs and manufactures the vast and complicated machinery: they lead two beautifully co-ordinated but distinct lives.[19]

These lives are divided, the writers explain, between 'Town and Country': London and Sussex. As this article suggests, Schreiber's professional identity was enhanced by her ability to achieve a 'balance' between work and home. From one perspective, this representation of the 'woman designer' at home could be offered as an example of the limited scope for women beyond the domestic environment.[20] However, this would be to overlook the importance of the home as a feature of representation also notable in the representation of her male contemporaries. For instance, the July 1955 issue of *US Vogue* featured an editorial, 'New Space Plan for Apartment Life', presenting Raymond Loewy's Manhattan apartment: 'beautiful, exciting and workable, it is a whole new way of planning space'.[21] The article featured a full-page photograph of Loewy's wife, 'pretty and dark haired', carrying their 2-year-old baby into the living room from the 'mobile-hung study' (see Colour Plate 10). 'Beyond it', the article continues, 'is a glimpse of Mr Loewy's workroom, with its Matisse lithographs from the book Jazz ... The Loewys spend six months of the year here, one in Palm Springs, the remaining in France.'[22] The portrayal is an excellent example of how the identity of the industrial designer rested on a glamorous and idealized image of work and home life; marital and domestic bliss worked to enhance the professional identity in this 'new profession'.

In Britain, as the previous chapter discussed, male Consultant Designers were also featured as personalities within the 'women's pages' of tabloid and broadsheet newspapers, as well as fashion and interiors magazines including *Homes and Gardens*, *Vogue* and *Tatler*. In 1958, *Homes and Gardens* magazine featured an article entitled 'The Home of an Artist and Designer in Pond Street, Hampstead', in which the writer focuses on the 'cleverly chosen colours' of the Henrion home, noting that 'the Henrions are experimenting all over the house with every kind of indoor plant from cactus to creeper'.[23] In this way, design offered a new proposition on the previously distinct boundaries between work and home life; labour and leisure. This was a highly gendered proposition, as it re-envisioned ideals about feminine and masculine identities at work.[24] Here, a well-balanced marriage, with the wife in the position of artist, functioned discursively to complement the serious, professionalized identity of the male designer. In March 1960, *Tatler* magazine published a double-page spread on Henrion's Hampstead home, paying particular attention to the gendered dimensions of this idealized work/living arrangement:

> Both the Henrions work at home, so she has a studio (off the garden) and he has two offices and a studio, a conversion done last year. His office is divided into working and reception areas by a long sofa (modern). Behind his desk (Danish teak) and chair (Victorian) is a wall of revolving bookshelves (his own design).[25]

This careful designation of the gendered work spaces within the home – the studio in the garden and the home office – marked out the practice and identity of design as a 'new profession', with new spatial arrangements that were modern and highly aspirational. In 1958, the British illustrated weekly publication *Sketch* featured an article entitled 'At Home with an Artist-Designer', introducing its upper-class readership to Richard Lonsdale-Hands, 'a very clever and prosperous artist and designer'.[26] The article paid particular attention to his home life, referencing his wife's status as an interior designer and presenting photographs of their bedroom alongside his home studio. Here again, the marital relationship proved an effective discursive tool through which to represent the 'in-between' status of design as both art and profession, drawing on feminine and masculine characteristics, as a 'modern' working identity forged somewhere in between the two.

Strikingly similar representational tactics were employed in the US, with reference to the design consultant and President of the Knoll furniture company Florence Knoll Bassett. Like Schreiber, Knoll maintained a tight grip on her archival footprint.[27] Meticulously collected press clippings have been preserved and digitized for permanent open access record by the Smithsonian Institution, securing the promotional value of her legacy well into the future.[28] An article in the *New York Times* in September 1964 described this 'woman who led an office revolution': 'The woman was young, dark eyed, dark haired and slender, with the clothes sense of a model, the training of an architect, unfailingly good judgement and the nickname of Sanu.'[29] Like Schreiber, Knoll was presented as an aspirational career woman and her marriage was an important part of this. The article talks through her 'working day': 'Lunch served on a tray by the pool, takes only half an hour. Then she works until 5:30 or 6, after which there are three sets of tennis and a swim with her husband.'[30] Here again, a blissful marital life is rendered complementary to the professional practice of design. Both Schreiber and Knoll Bassett capitalized on the exoticization of their European heritage in the media – and benefited from elite social networks secured through family connections, marriage and wealth.[31]

As these media representations imply, the roles of wife, housewife and 'woman designer' were interdependent and complementary. As 'experts' of married life, female designers could further extend their authority to the female consumer. In February 1952, Russel Wright, Freda Diamond and Dorothy Liebes spoke as experts at a 'Bridal School' run by the *Herald Tribune* to educate women on issues including fashion, design, human relations and food, conveying the new levels of expertise assigned to the role of housewife in the mid-century US. Diamond's marriage to industrial engineer Alfred Baruch was regularly commented upon as complementary to her professional identity. Diamond expressed this herself many times, stating that 'many women hold down jobs as well as being homemakers ... most husbands now enjoy a five day forty hour week and wives want to be free to spend their leisure time with them'.[32] An article in the *Sunday News* in February 1951 entitled 'Clever gal has design to thank for her living', focused on the gendered dynamics of her office and her marriage, presenting the two as compatible and complementary:

> After office hours, she becomes Mrs Alfred Baruch, wife of an industrial engineer. Alfred's office is on the main floor and the remainder of the building is home for this career couple. 'I think I can safely say that my marriage and career go well together,' Freda said. 'With a very understanding man to help me, I've combined the two for seventeen years.'[33]

Here, Diamond can be seen to utilize and instrumentalize her marriage as a device through which to claim authority and expertise over designing for home interiors. Diamond's 'feminine expertise' in fashion and homeware are positioned here as specialisms in balance with the technical proficiency of her husband's expertise as an engineer. An article in the *New York Times*, glowing on Diamond's impact on American household consumption and taste, also claimed that her husband checked her scale models for accuracy.[34] This image of a 'career couple' working together was an aspirational model for a new modern way of living and working, labour and leisure working in beautifully designed harmony.

The professionalization of industrial design in both Britain and the US took place in the feminized space of consumer culture.[35] Designers were highly conscious of the need to engage with feminine interests and concerns. They frequently identified the visitor to their World's Fair exhibits as 'she', and by the 1930s habitually referred to the American consumer as female.[36] In his 1930 career guide for industrial design, Harold Van Doren mused that while women were not 'naturally' inclined to design work, they were in fact 'closer students of trends than men. They spend hours shopping and talking to salespeople. They make excellent style scouts and astute buyers in the housewares departments of big stores.'[37] By the 1950s, the professionalization of the housewife was one outcome of advanced consumer culture in both Britain and the US.[38] In Britain, the CoID prioritized homemakers and housewives in their early promotional activities.[39] Misha Black was asked by the CoID to correct the copy for an educational booklet to accompany the 'Birth of an Egg Cup' section of the 'Britain Can Make It' exhibition, which would be sent to touring schools and Women's Institutes, and which compared the designer's use of decoration to the woman's use of make-up.[40] In an article in the *Los Angeles Herald Examiner* in October 1964, entitled 'He Knows About Women', Dreyfuss 'confessed' to learning 'domestic (and foreign) skills' such as cooking, cleaning and sewing. These 'feminine', domestic skills were 'permissible' within the hyper-masculine context of industrial design and complementary to the representation of the industrial designer as a 'new professional'.[41] Marketing and advertising agencies had long harnessed the values of 'feminine insight' by employing female psychologists and 'consumer engineers' to lead focus-group research projects for consumer products aimed at a female audience. As adman David Ogilvy put it, 'The consumer is not a moron: she is your wife.'[42]

Diamond, Schreiber and Knoll took advantage of their gendered identities as women to perform the roles of wife and designer – a duality that gave them advantages when facing the female consumer market. In June 1957, *The Times* presented a feature on Schreiber, 'Design for Air Travel', in which they reflected upon the value of feminine expertise in an increasingly competitive field: 'If women choose the airline on which they will travel, as they are said to choose the house a family buys, then there is value in a woman's outlook.'[43] The value of Schreiber's 'feminine expertise' was captured by her contemporary Consultant Designers in a birthday card they designed for her, in which she is depicted painting an aircraft pink (see Colour Plate 11).

A 'woman's eye' was said to be particularly valuable in relation to choices of interior design, including colour. The professional identity of 'woman designer' made these skills meaningful and valuable, so that femininity was not in conflict with professionalism, but rather a productive aspect of it. Freda Diamond perhaps most explicitly directed her expertise at the female consumer market. As she said in 1956:

> magazines and women's pages in newspapers have done a wonderful job educating the consumer and are constantly showing her what is new, beautiful and exciting, in good taste and in her price brackets ... the customer has changed. She is growing more style-conscious and is more articulate about what she wants ... Our homemaking consumer, for the most part is realistic, practical, intelligent and, as I said before, style-conscious.[44]

Diamond wrote consistently to defend the intelligence of the modern housewife in women's magazines, positioning herself professionally as an expert on women's fashion and taste. Her media presence, managed by her publicist Constance Hope,[45] worked to capture these qualities, presenting her as a 'designer for everybody' and 'a girl's best friend'.[46] Diamond also pointed out the advantages of working as a woman in the design profession. As she reflected in 1995:

> Russel Wright once asked me, 'Freda, how come I can only get a job from project to project and you can get people to sign up on a contract basis for years?' And I said, 'Russel, when I design something, it doesn't have to be what The Great Freda Diamond designs like it does with the Great Russel Wright. You're asking them to buy your name.'[47]

Here Diamond expresses a qualitative difference between the identity of male and female design roles, within the gendered context of post-war design consumption. As the previous two chapters argued, male design consultants in the US had been the loudest proponents of obsolescence, but her own relationship with the retail industry (and department stores in particular) was formed through a more strategic relationship based on trust and longevity. In 1959, she spoke at the Home Furnishings Market Conference against the value of obsolescence, which she presented as harmful to the profession because it was 'not serving the public': 'It's not what's new that's important, it is what's good', importing the European values of Good Design in this context.[48]

Publicising the profession

Alongside design, management consultancy and advertising, publicity was another of the 'new professions' that bloomed in the early years of consumer capitalism. Like design also, its history in the US is closely tied to that of the corporation. American advertising historian Roland Marchand documented the dramatic rise of public relations within his history of corporate identity in the US in the 1930s. As he states, PR consultants were afforded a new-found cultural status and respect within the corporation, whereas it had been 'a type of work once regarded almost with disrespect'.[49] As Chapter 1 suggested, the line between publicity and design work was blurred in the formative years of the profession in the US, as the New York-based design consultants, including Walter Dorwin Teague, instrumentalized publicity as a tool to enhance their professional status. By the middle of the century, publicity had come to occupy a sizeable portion of these 'big four' consultancies, a fact observed with some frustration by those working in other parts of the country.[50] Raymond Loewy, once characterized by George Nelson as the 'least publicized' in his *Fortune* magazine article in 1934, had gained a reputation for his exuberant media presence – in magazines, newspapers and also, increasingly, on television. Rumours abounded as to the amount of money he spent on publicity within his office, with one (unverified) estimate being $40,000 per year.[51] In taxes filed by industrial design consultant Egmont Arens for 1946, publicity and promotion was his single biggest office expense.[52] In an interview with Raymond Spilman, Chicago-based design consultant Dave Chapman reflected on his use of publicity, 'We had a program running those years, but it ran maybe $25,000, including the man we bought. We put the man on employment and he came in at about $4,000. And generated about that much.'[53] For a profession with limited governmental support and patronage, 'no consultant could afford to hide under a bushel' and publicity was considered a necessary factor of design work, to attract clients.[54] Increasingly, the role of publicity was also about building a 'corporate image' of the designer in the eyes of business, underlining the influence of the corporation at every turn in the professionalization of the industrial designer in the US.[55]

Publicity was regarded, in a broader sense, as an essential tool in professionalization. As Teague explained to the SID membership in 1945, 'books and magazine articles and all such informative material' are considered by 'legal experts' as 'evidence of the professional nature of the field'.[56] As such, his portfolio was presented in 1944 in the legal claim to overturn the Consultant Designer's tax status from non-professional to professional. This further explains the prevalence of the designer autobiography as a feature of the early years of professionalization in US industrial design, by all of the leading male protagonists.[57] Speaking to Raymond Spilman, Chapman remembered that Teague had employed the services of a professional PR consultant to 'help him write his book'.[58] Many of these texts were written in the form of memoirs as a kind of 'life writing', presenting the overlap between personal and professional that characterizes work in the precarious professions.[59] Indeed, professional design organizations, including the ADI and SID, made publicity a principal concern in their founding years. A list of no less than twenty-two 'national press members' in the 1945 brochure for the ADI reveals the scale of media operations that functioned as a platform for the industrial designer. These were both industry- and public-facing, and included *Better Homes and Gardens, Good*

Housekeeping, House and Garden, McCall's, Interiors, the *New York Times* and *The New Yorker*.[60] Many were ostensibly women's magazines, written for a female consumer and staffed by women, working as writers and administrators.

The SID in particular was anxious to establish some control over what they perceived to be a heavily 'sensationalized' picture of the industrial designer in public media.[61] Speaking at an early meeting of the Society in December 1945, Egmont Arens explained that publicity was not only very important for 'client attention', but also 'well-handled publicity for one designer benefits the whole field'.[62] Arens himself invested heavily in publicity within his own office. His wife, Camille, was also a staff writer at *McCall's*. As a 'new profession' that had grown in relation to both advertising and design, public relations offered significant opportunities for women to establish professional careers.[63] This was an area where 'feminine' expertise appeared to carry a professional advantage. In 1947, Louise Bonney Leicester was invited to speak as an expert on the subject of public relations and publicity to the SID on the subject of 'dignified publicity' for industrial design. Bonney Leicester, Director of the America at Home section of the New York World's Fair (1939–40) was thanked by Philip McConnell for her contribution as he acknowledged her 'great knowledge of the development of Industrial Design, long experience in public relations and her many contacts among key people in New York', highlighting the value of social contacts in this work. Bonney Leicester advised the Society against the use of advertising, not on principle, but because it would 'make it harder to place material in editorial and news' sections of the media, identifying the sophisticated strategic approach pursued by publicists in contrast with the less subtle and more direct appearance of the designer in corporate advertisements, described in the previous chapter.[64] This consultation represented a rare moment of deferment to female expertise by the all-male designers in attendance at this early SID meeting, on a subject very high on its agenda.

The role of women in the field of public relations was recognized by Ralph Caplan, Betsy Darrach and Ursula McHugh in *Industrial Design* in October 1960, in the editorial feature, 'Public Relations for a Profession', which provided a detailed overview of the relationship between the industrial designer and public relations, with interviews from those on both sides. The article reveals gendered attitudes to publicity and public relations, which was considered to be 'women's work' in the design office, as the authors suggested that the job 'can be handled even by an intelligent secretary … she doesn't even have to write, she can just call up the editor and whether she knows him or not is inconsequential'.[65] Even better, it suggests, designers can draw on the skills of a journalist wife (a situation utilized to great effect by designers Egmont Arens and Leon Gordon-Miller), while conceding that 'most designers do not have journalist wives and most want more sustained publicity than a secretary can perform in moments between her other duties'.[66] While the authors of the report define the role of press officer and publicity agent in male terms ('PR man'), it quickly becomes clear, through interviews with leading protagonists of the profession, that many active and successful press agents and publicists for industrial designers were female. Indeed, the origins of the role in the context of industrial design is attributed directly to one woman in particular, Betty Reese, publicist for Raymond Loewy Associates (RLA), pictured with Loewy in *ID* (see Colour Plate 8).

The descriptions of Reese in this feature echo something of the 'pioneering' language used to describe the first industrial design consultants. As they put it, 'As far as industrial design is concerned, there was not much activity that could truly be called public relations until 1941 when Raymond Loewy hired a blonde publicist named Betty Reese.' Reese established a reputation as

> one of the most highly respected business publicists in the nation and her success at RLA has had its effect on the way other designers think of publicity. One metropolitan office a few years ago wanted to hire a public relations director but refused to interview any men for the job on the grounds that 'Raymond Loewy has a woman and that seems to be what's needed.'[67]

Reese's reaction to this statement, as reported in the article, is powerful and revealing: 'It's not that I was a woman, but that I was a pro when I arrived at RLA.' Her self-confidence in her position within the RLA office and in her professional expertise is striking and visible in the commanding position she assumes in photographs of RLA board meetings.

Reflecting on her skills in the role, Reese emphasized the value of interpretation: 'I try to help the designer define what he is doing: the field has changed so much that he doesn't always have a chance to keep up with himself. But my main job is to interpret him to others.'[68] As this description suggests, the line between public relations, publicity and industrial design work was blurred. Publicity work had, by the mid-twentieth century, become so integral to the success of a designer's business, that it was integrated in the design process itself. Reese explained, 'working as a PR agent within a design consultancy was also considered a fringe benefit that would attract clients, since it would indirectly enhance the company's design program'.[69] In this way, publicity and press, key tools of professionalization in industrial design, constituted design work. As Sally Swing, executive secretary of the ASID put it, 'although 75 per cent of what a PR representative does for a designer can be done by almost anyone, the other 25 per cent is so close to design itself that it requires someone extraordinarily sensitive and able and usually expensive'.[70]

'As a new profession,' Reese stated, 'industrial design can be heaven for any press agent because it is so broad. There is no magazine in the

3.2 Betty Reese at a board meeting of Raymond Loewy Associates (n.d., c.1950). Photographer: unknown. Raymond Loewy archive, Cooper Hewitt Smithsonian Design Museum.

country for which you haven't got at least a story possibility.'[71] Reese was highly inventive in generating a buzz not only around Loewy as an individual, but around the persona of his workplace too, regularly sending press releases with humorous 'insider' stories on Loewy's birthday parties and other festivities, cultivating a culture of creative work, which constructed an image of the design office as an exciting and attractive workspace, quite at odds with some accounts of the intense labour conditions described by employees.[72] Reflecting on her relationship with Loewy at the time of his death, Reese said:

> Loewy was a great manipulator – one of the best – and he learned how to manipulate the public's image of himself. He decided at the age of about twelve or thirteen to create this character called Raymond Loewy and that's what he perfected ... Those were exciting years for me, because Loewy was the real thing. For a publicist, he was the greatest property, better than any star or celebrity. I had always wanted to be a stage director and with Loewy I had the perfect showman to direct.[73]

Reese regularly characterized her role as 'directing' Loewy's professional career, in a powerful statement of her agency within the RLA office. In later years, industrial design journalist and design critic Ralph Caplan reflected on the importance of Reese as a 'go-between' in securing a powerful media presence for Raymond Loewy. As he recounted to historian Alice Twemlow, Reese's role as intermediary in setting up a lengthy profile on Loewy's partner William Snaith went some way to restructuring the Loewy organization itself:

> When it was published, she came to me and said, 'I want to invite you to the Loewy Christmas party.' Now this was a party strictly for Loewy people only, no clients, nor press or anything. At the party Loewy announced that after all these years of refusing to share billing with anybody or anything, that they had decided to change the name of the organization from 'Loewy' to 'Loewy-Snaith.' Because, after that article, Snaith was able to say to Loewy, now everybody knows how the business works and what I do. I realised that the reason I'd been invited was to see what we had done. It made me think that what I write and publish has a consequence.[74]

While Caplan reflects here on his own agency as a design journalist and critic, it also emphasizes the centrality of Reese in the RLA office and its public and professional identity.

Reese gained a reputation within the architecture and design professions, but she maintained privacy in her personal life. In 1957, one of her Loewy colleagues, architect Andrew Geller, designed a beach house on Long Island seafront for Reese on the 'impossibly small budget of $5,000'.[75] Reese spent much of her professional life submitting press releases and placing editorial features in the architecture and lifestyle press for Loewy and his many holiday homes – including on Long Island. Nevertheless, while Reese worked to promote Geller's work, inviting her friends at the *New York Times* to see the property and have it featured on the newspaper's front page of its real-estate section, her name was never mentioned in the editorial as she maintained a strict professional separation from her work and her private life. In an interview with an architectural critic some years later, Geller described Reese as 'a strong-willed, independent career woman, who knew exactly what she wanted – intimate contact with the sea and instant release from her busy schedule in the city. She went about inventing her own style of life at the beach. The sleek and simple lines of the house captured her independent spirit and dynamic lifestyle.'[76]

Reese's status within the RLA office generated internal jealousy and criticism by designers employed within the consultancy office. A satirical newsletter, 'Life with Loewy' (see Figure 3.3), written and published by an anonymous staff designer and circulated internally, was especially harsh and misogynistic in its portrayal of Reese, referred to under the name 'B Smith Reese'. In May 1944, it featured a short article, 'Publicity and its Relationship to Industrial Design'. The article makes clear that the publicity department was not held in high regard by the design staff, claiming that it was 'written by request of Vice President Barnhart who

3.3 'Pin-up girl of the month' from *Life with Loewy*, 1:3 (1 May 1944). Raymond Loewy archive, Cooper Hewitt Smithsonian Design Museum. Betty Reese is pictured bottom right.

was reported as saying, "For Christ's Sake what does Reese do around here anyway?"' The author mocks the 'scope' of the publicity department, which had increased to such a scale that it was

> necessary for the body (clothed inadequately) to move to the Mexico City Offices for a quick promotion of Loewy interests ... which consumed five and a half weeks... At the outset of this BUSINESS trip, it was thought ... that the business of 'Stinkie' Reese (as she is known in the trade of course) could be concluded in two or three weeks. This was proven false ... since her suntan in two weeks was 'toast' and the trade is promoting 'burnt plastic shell' for the Fall line ... Mr Loewy cannot be reached for a statement. His remarks concerning the publicity department included the words 'done in' and 'done'. Still the scope of this department widens.

The author continues to criticize the 'variety' of the 'high klass' [sic] publicity. 'In *House and Garden* issue, six pages will be devoted to the house of Raymond Loewy in Mexico. *Architectural Forum* is planning a story on the house in Mexico. It can be expected that INTERIORS will use several photographs of the house in Mexico before the year and B Smith Reese, has grown much older. Mrs Loewy has been photographed on the terrace of the house in Mexico and can expect to see herself in several of the glossier fashion magazines, which incidentally are doing a great job of winning the war.'[77] This derogatory attack is accompanied by an image of Reese as 'Pin-up girl of the month', with lurid innuendos, describing how Reese, 'with her feet on the desk, receives the gentlemen of the press'.

The newsletter gives a revealing insight into the misogynistic culture that permeated and intersected with the 'creative' studio culture of the design consultancy. It also reveals some of the internal frustrations and tensions felt by employees within Loewy's office due to the high priority he placed on publicity within his organization. The context of war looms heavily in the newsletter and might have contributed to the heightened, aggressive culture of masculinity expressed in its pages.[78]

Managing the profession

The gendered dynamics of office politics were ingrained in the division of labour between men working in design and women, who provided supportive, 'backroom' administration. It was evident in the highly descriptive job roles within the US design consultancy, described in Chapter 2.[79] The wage brackets of receptionists and secretaries were among the lowest in the office, along with laboratory assistants. Likewise, the bulk of archival documents held within the archives of individual designers and their consultancies are composed of letters, financial administration and other administrative documents signed and filed by female employees. In oral histories conducted for the SIA and IDSA archives, female secretaries, assistants and office managers are frequently named by industrial designers – often in terms of high praise – but rarely form the subject of any interview, with rare exceptions.[80] Nevertheless, for some design organizations, including the first and most successful British design consultancy, the DRU, the administrative tasks of design, from library to business management, were explicitly acknowledged as an integrated part of the design service. As DRU employee J. Beresford-Evans put it in Italian industrial design magazine *Stile Industria* in 1958:

> Contracts, fees and such matters are largely dealt with by the business manager Dorothy Goslett. She, with the progressing officers and administrative staff, is also concerned with the organization of time schedules and the progressing of jobs. Research, reference and a large collection of materials and specifications is in charge of the librarian, so that all designers can quickly call upon reference materials and samples. These common services are comparatively expensive because of adequate staffing and the very full services available, but they permit fuller deployment of designers' skills, allowing them to concentrate on creative and productive work.[81]

As this statement suggests, the DRU put a value on administration as a function of design work. In this promotional piece on the renaming of the consultancy, Beresford-Evans emphasizes the flat organizational structure of the Unit, representing a new approach to work in a new sphere of professional practice. In their history of the DRU, John and Avril Blake commented upon the unusually prominent role Goslett carved out for administration within the DRU practice, which made up roughly one-third of the organization's employment,[82] stating, 'Office administration is usually considered to be a back-room job, but Dorothy Goslett has widened the context of her work by using her experience to write a guide for all embryo design offices.'[83] Her book, *The Professional Practice of Design*, communicated to designers that they must be idealistic and creative, but also 'practical business men' to turn their ideals into reality. Indeed, the authors suggest that Goslett's practical capabilities and good management acted as a core to the consultancy's reputation as the 'practical idealists'.[84]

Dorothy Goslett's centrality to the DRU office is indicated in the central position she assumes in the photograph shown in Figure 3.4.

Her skill in the role seems to have rested on her knowledge of the role of secretarial tasks (designated in the female gender throughout her book) such as filing, in which she expresses intimate knowledge of boxes, cupboards, labels ('tie on, *not* stick-on') and other everyday tasks of co-ordination, alongside costing, fees and business matters that ensured the successful running of a design consultancy office.[85] Although the book contains relatively scant information on the subject of publicity, it seems that Goslett was also in charge of this aspect of the DRU's work, and she gives some specific advice in the book about the maintenance of the press-clippings folder.[86] Goslett's professional history gave her a good

3.4 Dorothy Goslett at the centre of the Design Research Unit (1946). Photographer: unknown. Misha Black Archive, V&A Archive of Art and Design, AAD/1980/3/7

basis from which to perform the role of a publicist, since, as she states in an interview with CSD president Robert Wetmore, she started work as a typist and secretary for a small magazine called *The Needlewoman*, to producing the Harrods Newsletter, to joining an advertising agency as a secretary and then moving into the Ministry of Information during the war, where she got to know and work closely with Milner Gray and Misha Black while organizing war-propaganda campaigns. Here, Goslett states, she built 'a reputation for organizing things', a descriptor commonly apportioned to women's work within design and other professions.

The scope of Goslett's role within the Consultancy was clearly immense, as she tells Robert Wetmore, 'I was business manager, right from the start. I had all the staff problems and all the money problems and general administration of it all.'[87] In this interview, she also takes pride in remembering her role in shaping the consultancy's non-commercial values, agreeing with Misha Black to allow architects to take on work which she knew would lose money, but give them 'happy architects'.[88] Throughout this interview, Goslett explicitly gendered work within the DRU office, in which the male designers and architects were named, but the female staff ('girls'), remain anonymized.[89] It is clear therefore that while there may have been an open and democratic working structure within the organization that claimed to give an equal voice to its male employees, this was not a privilege afforded to its female administrative staff.

Goslett was encouraged, during a DRU meeting, by furniture and interior designer Ernest Race to 'write a standard book to tell young designers how to behave and what to do'.[90] The resulting publication, *The Professional Practice of Design*, was published to general acclaim within the profession in 1960, although *Design* magazine noted the 'partisan' nature of its advice, which reiterated verbatim the professional values and ethics of the SIA. The book was published four times (1960, 1971, 1978 and 2004), suggesting a wide and consistent readership of designers eager for practical guidance. In 2004, an exchange of letters about the book's republication in *Design Week* identified divided attitudes about the book's utility for the contemporary designer. While graphic designer and art director Adrian Shaughnessy admired the practicality and clarity of its writing, calling it a 'marvellous book, full of sage and timeless advice', others critiqued its misogyny and narrow conception of the practice of design work.[91] Indeed, the practical and commonsensical structure and language of the book stand in marked contrast to career guidance manuals for the profession published in the US, including Harold Van Doren's widely cited *Industrial Design: A Practical Guide* (1940) and Arthur J. Pulos's *Opportunities for Careers in Industrial Design* (1970), which was published only one year earlier. While Pulos and Van Doren both place a great emphasis on defining the importance and status of the designer in industry, his character (all three texts are written in the male gender) and his training, neither can match the practicality of Goslett's writing, which provides, in great detail, all the information required for setting up an office in design. Moreover, while Van Doren and Pulos, both designers themselves, champion the greatness of the industrial designer on a grand and expansive scale ('the industrial designer, because of his broad and unprejudiced knowledge, and because of his innate sensitivity to his environment, is privileged to stand at the threshold of tomorrow and because of his keener vision and richer imagination is able to sense what lies beyond'),[92] Goslett's writing has a much more deprecating view ('one of those unpractical, undependable artist chaps – long-haired, unshaven, corduroy-trousered, sandal-shod, generally unkempt, probably amoral, thinks he knows everything').[93] Goslett's impression of the designer was clearly underpinned by a more critical impression of the designer's performance of hyper-masculine status. From her perspective, professionalism was a tool through which to turn this privileged expertise into something useful and practical.

Nevertheless, Goslett was by no means a progressive advocate for women in the industry and was regularly one of the most conservative voices within SIA meetings on the question of professionalism in design, as discussed in the next chapter. The practical nature of her book gives a sobering insight into the limited space in which she and other women working in the field were expected to conduct their professional lives.

For instance, in a section on 'Finding Clients', she writes of how entertaining is a 'must for professional man who wants to build up his practice. The little rituals of pouring a drink, choosing a meal, initial small talk and lighting cigarettes all help', confirming the gentlemanly ideals of masculinity to which the British designer conformed. Goslett acknowledged that this culture of entertainment was not aligned to feminine gender etiquette, stating, 'the woman designer has one problem facing her as a hostess which over-rides all others. No male guest is going to like it if she pays the bill or is even going to let her.' Goslett's solution, ever the pragmatist, was to join a gentleman's club or use a credit card. 'The second alternative will probably be the best since most women's clubs tend to have a rather cloistered atmosphere and even the dining room for mixed guests may be overwhelmingly and tweedily feminine', she adds.[94] These were hardly motivational or inspiring words for aspirant female designers reading the book.

It is worth briefly pausing to consider the distinctive ways in which Goslett's role as manager within the DRU has been accounted for in contrast to Michael Farr, 'pioneer' of design management in Britain. As discussed in the previous chapter, Farr's book, *Design Management*, published in 1966, represented a transition for the design profession as design came to be represented within business as a 'unique factor in competition'. Formulated as a scientific method under the principles of the Design Methods movement to which Farr was allied, Design Management articulated a new, highly specialized role within the profession. This was a position definitively presented by Farr as a male form of expertise. As he put it, 'His job – in brief – is to investigate from the designing point of view the requirements for a new product; find and brief the designer (or team of designers); set up and operate an easily understood network of communication between all parties concerned in the new product and be responsible for the co-ordination of the project until the prototype reaches the production line.'[95] While much of the substance of Farr's role could be understood as identical in scope to the tasks undertaken by Goslett in her role within the DRU, Farr's depiction of the role and his structured, scientific approach clearly designated it as a masculine practice.

Professional gatekeepers

Photographs of female employees, dutifully seated behind typewriters in design offices, are a regular feature within the archives of design organizations in Britain and the US, presenting a static and relatively inactive role that betrays the reality of their agency. In 1945, the Council of Industrial Design (CoID) assumed responsibility for the Record of Designers, taking over from the independent public institution, the NRIAD. In his oral history, contributed to the CSD archive in 1984, designer David Harris told Robert Wetmore:

> I had ambitions like everyone else at this time. I went along to the Council of Industrial Design and two charming ladies were in control – Miss Lomas and Miss Tomrley – and I was very reticent to go up those stairs and see them in Petty France as a young sort of twenty-four-year-old … because I knew that they were placing for men like Eames and another man called Henrion.[96]

Indeed, throughout the interviews conducted by Wetmore for this oral history archive, 'Miss Lomas and Miss Tomrley' are frequently positioned in this 'gatekeeping' role for designers in Britain remembering the beginning of their professional careers. Tomrley was employed by the CoID less than a year after the Council was founded and is referred to in the First Annual Report (1945–1946) as primarily responsible for 'Design Advice/List of Designers' initiatives and also secretary assigned to the Design committee.[97] She was a highly trusted member of the Council's Senior Staff, involved from the earliest years in committees including the Advisory Committee on the Design of Consumer Goods in 1946–1947 and was assigned responsibility in the engagement with naturalization of foreign designers, the demobilization of designers still in the services and aspects of Design Training.[98] By 1949, she was signing her letters 'Senior Design Officer'.[99] Administrative files reveal that she was employed part-time, though had become full-time by the end of her years at the Council,[100] at Senior Officer Grade III, a level occupied by six employees, four of whom were female.[101] These records show that male administrative employees enjoyed greater mobility through the pay scales, whereas no female employee appears to have moved above III, Tomrley's grade.[102]

Although the Record was described as an objective and neutral point of contact between designer and industry, Tomrley clearly applied value-based judgements in her management of it. Only one surviving folder of recommendations of the Record survives in the Council's archives. However, a fuller picture of the system by which Tomrley managed and administered the Record can be pieced together through research in other areas of the Council's archive, including a collection of organizational material for the 1956 exhibition, 'Designers At Work' and a report entitled *Student Behaviour*. The 'Designers at Work' exhibition was organized in association with the SIA and was the first exhibition staged at the CoID's new Design Centre. This was to be a 'special display … showing in actual samples and photographs, the results of recommendations made by the Council of Industrial Design'.[103] While the exhibition aimed to present the Record as a neutral point of contact between designer and industry, correspondence between Tomrley and the Council's Exhibitions Manager Philip Fellowes shows that this was not the case:

> I have taken our large concertina files and I have divided them into sheep and goats. If you take the folder with the pink guide cards, these are the sheep. I do not think you will need to supplement them with the goats.[104]

This documentation reveals Tomrley's curatorial role in selecting and managing the Record and its representation in public. It also forms some useful context from which to read the report *Student Behaviour*, published one year later.

Tomrley drafted the document entitled *Student Behaviour* in 1957, addressed to Robin Darwin, principal of the Royal College of Art (RCA) to observe a shift in the attitude, physical appearance, manner and behaviour of RCA male graduates. Tomrley claimed that she had been prompted to write it because of a small number of complaints from industrialists who had interviewed RCA students, including A. Gardner Medwin, who wrote, 'What students lack is what is known in industry as "personal qualities".'[105] This referred to matters of physical appearance and dress, tidiness, manners and professional behaviour. Nevertheless, the emotionally charged language of the report, coupled with the fact that it was directed 'entirely at the men: the girls have much higher standards',[106] does suggest that, like Dorothy Goslett, Lomas and Tomrley were personally motivated by their impression of a culture of entitlement and privilege within the male student graduates at the RCA, which they perceived to be both arrogant and slovenly. She wrote that she and Miss Lomas had been 'disturbed' by the 'bad behaviour' of students, which fell under two general headings:

> a) little interest in the traditions, attitudes and necessities of the business world and a general air of conferring a favour by talking to businessmen.
> b) Untidy and unsuitable dress, dirty hands, lack of grooming about the head, unpunctuality and slovenliness in answering letters and casual way of presenting drawings.[107]

The report gives specific examples of these 'delinquent males'.[108] They included Eddie Pond, J. V. Sharp, W. L. Belcher, N. Morgan, R. Atkins and K. Lessons. Of Pond, it was said, 'Mr Crawford found his manner, bearing, grooming and cleanliness so deficient that he felt unable to commission work; his person and the casual presentation of his portfolio created an impression that he would be casual in all matters.'[109]

In a series of letters between RCA Principal Robin Darwin and Tomrley, Tomrley hints towards the privilege of the RCA student, stating, 'We have excellent reason for being more concerned over the RCA than any other school … It is still the only school to which Miss Lomas and I give two or more whole days of our time to interview twenty to thirty students with their work.'[110] Trying to endear Darwin to her perspective, Tomrley explained how she understood that 'most of the students come from the provinces and are in London for the first time', 'many have never left home before', and were 'surviving on small grants and living in poor accommodation where it is difficult to maintain high standards of hygiene'. 'The whole situation adds up to a removal of all forms of discipline, other than that of their designing, just when they most need tightening up', she added.[111] This attitude echoed earlier government reports, which proposed an 'environmental reading of creativity'.[112] For example in 1944, the Dress Committee of the Council for Art and Industry argued that 'The designer's chief handicap is his provincial environment. His sense of style and fashion is conditioned by what he sees in his native town.'[113]

Organizing the profession

While, as Chapter 1 noted, female membership of the ADI was a minority component, women were highly active within the Institute and central to its organization and promotion before and during the war. In March 1945, Egmont Arens wrote to SID secretary Philip McConnell to say that Ruth Gerth, a 'born organizer', had worked very hard on the Artists Guild and other artists' organizations and it was 'largely due to her efforts that the New York chapter of the ADI has been successful'.[114] Nevertheless, her membership is listed under her husband's name as they worked together in partnership in the firm Kosmac + Gerth.[115] This representation of women as 'born organizers' and administrators has a longer social history, and the design profession was no exception.[116] Even within the ADI, in which women were relatively more active than in the SID, their roles were often administrative and organizational. In 1960, Leon Gordon Miller wrote to the executive committee members of the IDI to say that it was traditional to give a Christmas gift to the Executive Secretary; 'perhaps Ann Franke can do the shopping for us'.[117] Franke, an industrial design consultant, had been pivotal to the formation of the New York chapter for the Society. Design organizations in Britain and the US were relatively constrained by the misogynistic culture that governed societies in both countries. Their 'professional histories' reveal an essentially 'supportive' view of the female contribution to the Society's activities and identity. In his history of the CSD for instance, published in 1980, James Holland remarks upon the loyalty of secretaries and wives:

> Nor have Presidential wives failed to make their contribution to the frequent occasions when they have been required to stand at the presidential left hand, and they have kept their diaries assiduously and encountered reception lines while sustaining welcoming smiles.[118]

A similar representation of the role played by women within the SID is found in documentation for its annual meetings. The SID's Annual Conference took place at the Playboy Hotel on Lake Geneva, Wisconsin at least twice, and invitations were extended to 'member's wives', a convention that was also common practice in Britain and continued in the conventions of the International Design Conference in Aspen (IDCA) in the 1960s and 1970s.[119] This gendered language, which put women in the position of 'wife', was common even in professional organizations where the female representation was higher. Ladies were graciously invited to the meetings of the American Ceramics Association, for instance, to 'give them some idea of the problems, headaches which their executives, engineer and designer husbands face'.[120] Wives were encouraged to attend the SID's annual meetings, and a full programme of events, arranged by the 'Wives' Coordinator', which included a 'famous fashion speaker, round-table discussion on exotic foods, jewelry' and – intriguingly – 'living with designers'. This usually included a shopping excursion. A letter from Eugene Gerbreux, SID secretary, confirmed this method had helped to increase the attendance figures.[121]

The merger of the IDI with the more exclusively male-dominated SID had a negative impact on the representation of women within the newly merged IDSA in 1965. The IDI membership directory for 1962 showed nine female Consultant Designers, but these women seem to disappear from IDSA directories after the merger.[122] In 1964, Belle Kogan wrote to the Board of Directors as Chair of the nominating committee for the New York chapter of IDI, confirming their nomination of Miss Elizabeth Dralle, who was 'dedicated to the service of IDI for twenty + years, an articulate and respected member of the design profession, who gave devotion and service to IDI'. She added, 'this would be the time to give her this recognition. Impending changes in the organization may make this impossible in the future', referring to the impending merger between the IDI and SID.[123] Dralle had joined the ADI in 1942, serving as secretary to New York chapter in 1945, and wrote the important 'IDI and You' document circulated among students to recruit them to the Society. She was then National Secretary, had trained the next National Secretary and was key in fostering a friendship between the ASID and IDI. Dralle received no votes from the Board and was not elected to Fellow. The list of new members for IDSA in 1965 includes no women.

Conclusion

Design history has often privileged the role played by pioneering male designers, but, as this chapter has shown, this history has obscured the complex dynamics of gender at work through the process of professionalization, which involved a negotiation between masculine and feminine identities. As a 'new profession' forged between production and consumption, the identity of the industrial designer interacted with and depended upon women acting in diverse roles and contexts; as consumers, housewives, administrators, organizers, managers and, in some cases, designers. A considerably richer and more complex account of the process of professionalization thereby emerges by opening up the parameters of professional work to include promotional, administrative, organizational and management roles within design.

Gendered analysis of the professionalization of industrial design illuminates fascinating ambiguities between work and leisure in the representation of design as a 'new profession'. For instance, the co-constructed identity of the 'woman designer', as seen in the case studies of Schreiber, Knoll and Diamond, was strengthened through their positive engagement with 'feminine' attributes including beauty, glamour, taste and their social marital status as wives and homemakers. There is much more to unpack here than has been possible within the scope of this chapter on the boundaries of these concepts and the social, racial and class-based hierarchies on which they were built and performed. The subject of marriage opens itself up in surprisingly rich and complex ways in relation to the representation of the designer. Traditionally, much work on the many 'married partnerships' in twentieth-century design practice – a relatively common phenomenon that shaped the working identities and media representations of designers from Jacqueline Groag, Lucienne Day, Florence Knoll Bassett and Ray Eames, among many others – has looked at the 'problem' of marriage as another way in which women's careers were 'hidden' from public view and behind the identity of their husbands. However, as this chapter has shown, marriage functioned as a representational tool through which male and female design consultants in the post-war period connected to their consumer – the housewife. In particular, the marriage between an interior designer or artist (female) and industrial designer (male) was one of the ways in which 'gender balance' was achieved in the mediation of the profession to the public.

Looking beyond the individual male or female designer, women played a central role in other professional spaces, including publicity, administration and organization, relatively undocumented sites of professional practice. This distributed view of agency in the profession exposes the limitations of focusing on the designer-auteur when accounting for the professionalization of the field, contributing to emergent work on the 'unknown woman' in design history and the making of built space.[124] Publicity, it has shown, in particular, was instrumentalized by designers in the US and, increasingly, in Britain, where visibility was regarded as an essential tool in professionalization. As a focused study of Betty Reese's role has revealed, the lines between publicity and design were especially blurred in the context of US industrial design, where the designer depended upon visibility in the media for patronage. Reese's power over the Loewy brand identity upturns the agency of the individual designer and reveals the hidden and unseen mechanisms at work behind the illusion of professional identity.[125]

As Katarina Serulus has shown, in her fascinating research on the role played by Josine des Cressonnières through the International Council of Societies of Industrial Design (ICSID), femininity could be mobilized as an effective tool in the diplomatic performance of design on the international stage.[126] Here, des Cressonnières's role extended well beyond the limited public role offered to women in positions of power within the SID or SIA, putting in perspective the regressive impact of professionalization within national organizations for design, in contrast with the opening-up of a more dynamic international scene. While archival documentation for women working in administrative and organizational functions within these national organizations is frustratingly sparse, the evidence available allows us to speculate on how Dorothy Goslett and Cycill Tomrley both framed their attitudes to professionalization in relation to a perception of unruly male arrogance. For them, professional manners and behaviour could be used to regulate and govern the egotistical performance of hyper-masculinity they worked with. Continuing with this theme, the next chapter looks more closely at the behaviours and ideals that shaped and gave meaning to the profession, through the lens of the professional organization.

Notes

1. Charles Whitney, Opening of three-week exhibit, First Chicago Area Industrial Design Exhibition (11–30 October 1954), Illinois Institute of Technology, quoted in IDSA 107 Scrapbook, IDSA Archive, SCRC.
2. Harold Van Doren, *Industrial Design: A Practical Guide* (New York, McGraw Hill, 1940), pp. 80–1.
3. Van Doren makes this argument in *Industrial Design*.
4. Alice Twemlow, *Sifting the Trash: A History of Design Criticism* (Cambridge, MA: MIT Press, 2017), p. 55.
5. Ibid.
6. Important exceptions include Lynne Walker, *Women Architects: Their Work* (London: Sorella Press, 1984), Cheryl Buckley, *Potters and Paintresses: Women Designers in the Pottery Industry, 1870–1955* (London: The Women's Press, 1990) and Joan Rothschild (ed.), *Design and Feminism: Re-visioning Spaces, Places, and Everyday Things* (New Brunswick, NJ: Rutgers University Press, 1999). For a useful review of the field, see Judith Attfield, 'Review Essay: What Does History Have to Do With It? Feminism and Design History', *Journal of Design History*, 16:1 (2003), 77–87.
7. Elizabeth Darling and Lesley Whitworth, 'Introduction: Making Space and Remaking History', in Darling and Whitworth (eds), *Women and the Making of Built Space in England, 1870–1950* (London: Routledge, 2007), p. 2.
8. Andrea Komlosy, *Work: The Last 1,000 Years* (London: Verso, 2018).
9. Fiona MacCarthy to Gaby Schreiber (30 September 1988), Gaby Schreiber Archive, VAAD. June Fraser interview with Robert Wetmore (15 May 1985), CSDA.
10. See Jill Seddon, 'Mentioned but Denied Significance: Women Designers and the "Professionalization" of Design in Britain c. 1920–1951', *Gender and History*, 12:2 (2000), 426–7. These phrases continue to be adopted within the literature – see more recently, Pat Kirkham, *Women Designers in the USA 1900–2000: Diversity and Difference* (New Haven, CT: Yale University Press, 2000), Anne Massey, *Women in Design* (London: Thames & Hudson, 2022). 'Portraits: Women Designers' at the Fashion and Textile Museum, London (March–July 2012); see Leah Armstrong, http://arts.brighton.ac.uk/collections/design-archives/projects/women-designers, accessed 15 April 2024.
11. New names listed for that year included Beatrice Adams, Vice-President of Gardner Advertising; Maria Bergson, Industrial Designer and Architect; and 'Dress Designers' Freda Diamond, Anne Fogarty and Cecil Chapman, *Who's Who 1956*, Freda Diamond Collection, Box 1, Folder 18, NMAH.

12 Carroll Pursell, '"Am I a Lady or an Engineer?" The Origins of the Women's Engineering Society in Britain, 1918–1940', *Technology and Culture*, 34:1 (1993), 78–97.
13 Liz McQuiston, *Women in Design: A Contemporary View* (New York: Rizzoli, 1988), pp. 110–13.
14 Catherine Moriarty, 'A Backroom Service? The Photographic Library of the Council of Industrial Design, 1945–1965', *Journal of Design History*, 13:1 (2000), 39–57.
15 Bedell was Public Relations Director at Lippincott & Margulies.
16 Russell Miller, 'Gaby Schreiber's British Empire', *Harper's Bazaar* (July/August 1968), pp. 42–5, VAAD, AAD/1991/11/14/15.
17 Arthur Becvar, General Electric (20 June 1962), Spilman Papers, SCRC.
18 *Evening News* (8 November 1951), Gaby Schreiber Archive, AAD/2009/7/6.
19 'Town and Country, How Two Designers Enjoy a Life Combining the Best of Both', *House and Garden* (1957), Gaby Schreiber, VAAD.
20 See Armstrong, http://arts.brighton.ac.uk.
21 US *Vogue* (July 1955), Raymond Loewy Archive, CHMA.
22 Ibid.
23 'The home of an Artist and Designer in Pond Street, Hampstead', *Homes and Gardens*, 53 (1958), FHK Henrion, press clippings, UBDA.
24 See also 'Working from Home: Fashioning the Professional Designer in Britain', in Guy Julier et al. (eds), *Design Culture: Objects and Approaches* (London: Bloomsbury, 2019), pp. 131–45.
25 *Tatler* (March 1960), FHK Henrion, press clippings, UBDA.
26 'At home with an artist and designer', *Sketch* (1 January 1958), Richard Lonsdale-Hands, press clippings, HAG.
27 Indeed, Knoll's archival filing system bears a striking resemblance to Schreiber's, as both women used initial-embossed spiral-bound leather books to record press clippings and other ephemeral material, presented in the form of a 'family album' to document professional activities.
28 See Jennifer Kaufman-Buhler, 'Review: *No Compromise: The Work of Florence Knoll*', *Journal of Design History*, 35:2 (2022), 201–2.
29 'Woman who led an office revolution rules an empire of modern design', *New York Times* (1 September 1964), Florence Knoll Bassett Archive, digitized collection, NMAH.
30 Ibid.
31 Schreiber's first work in Britain was interior design for Lord Antrim. Her elite social connections in Britain are well documented in her scrapbooks and photograph albums. Knoll benefited from an elite design heritage. See Ana Araujo, *No Compromise: The Work of Florence Knoll* (Princeton, NJ: Princeton University Press, 2021).
32 Freda Diamond, 'Today's consumer ... she's more style conscious', Tufted Textile Manufacturers Association Directory 1956, p. 114, Freda Diamond Archive, NMAH.
33 'Clever gal has design to thank her for her living', *Sunday News* (25 February 1951), Freda Diamond, press clippings, NMAH.
34 Unknown author, 'Woman Designer Credits Husband For Her Success', *New York Times*, n.d. (c.1951–1956), Freda Diamond, press clippings, NMAH.
35 See Christopher Breward, *The Hidden Consumer: Masculinities, Fashion and City Life* (Manchester: Manchester University Press, 1999), Erica Rappaport, *Shopping for Pleasure: Women in the Making of London's West End* (Princeton, NJ: Princeton University Press, 2000), Colin Campbell, *The Romantic Ethic and the Spirit of Modern Consumerism* (Oxford: Blackwell, 1987).
36 Roland Marchand, 'The Designers Go to the Fair II: Norman Bel Geddes, The General Motors "Futurama," and the Visit to the Factory Transformed', *Design Issues*, 8:2 (Spring 1992), 281.

37 Van Doren, *Industrial Design*, p. 81.
38 Rachel Ritchie, 'The Housewife and the Modern: The Home and Appearance in Women's Magazines, 1954–1969' (PhD thesis, University of Manchester, UK, 2010).
39 See Lesley Whitworth, 'The Housewives' Committee of the Council of Industrial Design: A Brief Episode of Domestic Reconnoitring', in Darling and Whitworth (eds), *Women and the Making of Built Space in England*, pp. 181–97.
40 This statement has already been made in Chapter 1; see note 61: Mr Ironside to Mr Jarvis with corrections to 'planned copy for "birth of an egg cup"'. Ironside made a correction to page 15, 'He uses decoration as a woman uses make-up', 'This may, I'm sorry to say, be objectionable to some teachers and should be avoided' (30 June 1947), Misha Black Egg Cup Folder, Design Council Archive, UBDA.
41 See Joseph McBrinn, *Queering the Subversive Stitch: Men and the Culture of Needlework* (London: Bloomsbury, 2021, pp. 180–96.
42 Ogilvy, quoted in James Pilditch, *The Silent Salesman: How to Develop Packaging that Sells* (London: Business Books, 1973), p. 4.
43 Schreiber, Blue and Red Photo Albums, AAD/2009/7/6; AAD/2009/7/7; AAD/2009/7/8/1; AAD/2009/7/8/2, VAAD.
44 Diamond, 'Today's consumer', pp. 112–13.
45 Constance Hope, founder of Constance Hope Associates, was a highly successful press agent for artists and musicians, with clients including Fannie Hurst, Lotte Lehmann, Erich Leinsdorf, Lily Pons, Bruno Walter, Grace Moore, Leopold Stokowski and Leonard Bernstein. She was also the author of *Publicity is Broccoli* (New York: Bobbs Merrill, 1941). Her papers (1931–1975) are held at the Columbia University Archive, New York.
46 'Designer for Everybody', *Life* (4 April 1954).
47 Freda Diamond, quoted in Mike Musselman, 'Designing Woman', unknown publication, January/February 1995. Freda Diamond Archive, NMAH.
48 'What's Good, Not What's New Aim of Furniture Designers', *Gazette* (18 June 1959), Freda Diamond, press clippings, NMAH.
49 Roland Marchand, *Creating the Corporate Soul: The Rise of Public Relations and Corporate Imagery in American Big Business* (Berkeley: University of California Press), p. 203.
50 Raymond Spilman, in an interview with Dave Chapman (c.1968), Spilman Archive, SCRC.
51 In the mid-1950s, this was a significant amount. Ibid.
52 The amount claimed was $3,039.86. Egmont Arens, 'Tax', Folder 23 (29 November 1946), Arens papers, SCRC.
53 Raymond Spilman, interview with Dave Chapman, Spilman Archive, SCRC.
54 Walter Dorwin Teague to Raymond Spilman (4 May 1960), IDSA 62_9_005, IDSA Archive, Syracuse.
55 Raymond Spilman said this in his interview with Dave Chapman.
56 Walter Dorwin Teague, SID meeting, 15 October 1946, IDSA Archive, IDSA_2, SCRC.
57 Norman Bel Geddes's *Horizons* was published in 1932, Walter Dorwin Teague's *Design This Day* in 1949 and Raymond Loewy's *Never Leave Well Enough Alone* in 1951. Henry Dreyfuss, *Designing for People* (New York: Simon & Schuster, 1955).
58 Ray Spilman, interview with Dave Chapman.
59 See Heidi Egginton and Zoë Thomas (eds), *Precarious Professionals: Gender, Identities and Social Change in Modern Britain* (Chicago: University of Chicago Press, 2021).
60 *American Home, Architectural Forum, Architectural Record, Better Homes and Gardens, Good Housekeeping, Graphic Architecture, Home Furnishing Merchandising, House + Garden, House Beautiful, Industrial Publications, Interior Design &*

Decoration, Interiors, McCall's, Mademoiselle, National Furniture Review, New York Times, The New Yorker, Parents Magazine, Popular Home, Retailing, Women's Home Companion, ADI Membership brochure, 1945, John Vassos Archive, Syracuse.

61 SID meeting (17 December 1945), IDSA Archive, SCRC.
62 Ibid.
63 Historians of public relations and publicity work have recently started to correct the narrative of 'male pioneers' of the profession, in recognition of the industry's status as 'female-dominated'; see Anastasious Theofilou (ed.), *Women in PR History* (London: Routledge, 2021).
64 Louise Bonney Leicester, SID meeting (17 December 1945), Walter Dorwin Teague files, SCRC.
65 Ralph Caplan, Betsy Darrach and Ursula McHugh, 'Publicity for a Profession', *ID Magazine* (October 1960), pp. 74–85. Raymond Loewy Archive, CHMA.
66 Ibid.
67 Ibid.
68 Ibid.
69 Ibid.
70 Ibid.
71 Ibid.
72 For instance, she wrote a press release sent to the *New York Post* about the design specifications of Loewy's birthday cake (1 September 1963), Raymond Loewy Archive, CHMA.
73 Betty Reese, 'Profile', *Industrial Design* (December/January 1986), p. 41.
74 Alice Twemlow, interview with Ralph Caplan (2007), https://alicetwemlow.com/the-consequences-of-criticism-an-interview-with-ralph-caplan, accessed 4 April 2023.
75 Alastair Gordon, *Andrew Geller: Beach Houses* (Princeton, NJ: Princeton University Press, 2003), p. 23.
76 Ibid.
77 Newsletter, 'Life with Loewy' (1 May 1944), Raymond Loewy Archive, CHMRB.
78 See for instance Anne-Dorte Christensen and Palle Rasmussen (eds), *Masculinity, War and Violence* (London: Routledge, 2018).
79 See Chapter 2, p. 69.
80 For instance, Ed Kimmell, manager at the Walter Dorwin Teague Associates office, references Marion Johnson, a formidable secretary at the WDTA office, whom he refers to as 'the boss' and 'a rough little gal, quite a gal'. Ray Spilman interview with Ed Kimmell, *c.*1968, IDSA Archive, SCRC.
81 J. Beresford-Evans, *Stile Industria* (21 July 1958), Misha Black Archive, VAAD, AAD/1980/3/6.
82 'Of the 60 staff employed by DRU in 1967, one-third were concerned with clerical and administrative activities', John and Avril Blake, DRU: *The Practical Idealists: Twenty-Five Years of Designing for Industry* (London: Lund Humphries, 1969), pp. 62–3.
83 Ibid., p. 24.
84 Ibid.
85 Dorothy Goslett, *The Professional Practice of Design* (London: Batsford, 1960), p. 171.
86 Ibid., pp. 124–5.
87 Dorothy Goslett, interview with Robert Wetmore (23 January 1983), CSDA.
88 Ibid.
89 She remembers a 'big room in front and two girls in a back room, one doing accounts and one filing. Upstairs, Kenneth Bayes worked with one other girl', Goslett, interview with Robert Wetmore, CSDA.
90 Ibid.

91 Adrian O'Shaughnessy, 'Advice Worth Heeding', *Design Week* (5 August 2004), www.designweek.co.uk/issues/5-august-2004/advice-worth-heeding, accessed 15 April 2024.
92 Arthur J. Pulos, *Industrial Design Careers* (n.p.: Vocational Guidance Manuals, 1970), p. 155.
93 Goslett, *Professional Practice of Design*, p. 163.
94 Ibid., p. 36.
95 Michael Farr, 'Design Management: Why Is It Needed Now?', *Design* (August 1965), pp. 38–9.
96 David Harris, interview with Robert Wetmore (14 April 1984), Box 38, CSDA.
97 First Annual Report, Council of Industrial Design (1945), Design Council Archive, UBDA.
98 File: National Register of Industrial Art Designers, DCA/12/43, UBDA.
99 Letter, Cycill Tomrley (1949), DCA/14B/27/42.
100 File: 'establishment section' (2 December 1963), Design Council Archive, UBDA.
101 File 'establishment section' (1 March 1954), Design Council Archive, UBDA.
102 I am grateful to Lesley Whitworth, Curatorial Director, UBDA, for verifying this research.
103 Record of Designers Exhibition, Policy and Selection of Exhibits, Design Council Archive, UBDA.
104 This reference to 'separating the sheep from the goats' is biblical in origin and refers to the distinction of 'good from bad', referencing the popular maxim of 'Good Design' in mid-century British design reform. C. G. Tomrley to Philip Fellowes (17 July 1956), Record of Designers Exhibition, Policy and Selection of Exhibits, Design Council Archive, UBDA.
105 Tomrley, 'Student Behaviour' (1957), Design Council Archive, UBDA. See also Leah Armstrong, 'A New Image for a New Profession: Self-image and Representation in the Professionalization of Design in Britain, 1945–1960', *Journal of Consumer Culture*, 19:1 (2017), 104–24.
106 Ibid.
107 Ibid.
108 Ibid.
109 Ibid.
110 Cycill Tomrley to Robin Darwin and Mr Medwin (26 February 1957), UBDA.
111 Ibid.
112 Patrick Maguire and Jonathan Woodham (eds), *Design and Cultural Politics in Post-War Britain*: *The* Britain Can Make It *Exhibition of 1946* (London: Leicester University Press, 1998), pp. 112–22.
113 Ibid.
114 Egmont Arens to Philip McConnell, March 1945, IDSA Archive, Board of Directors, SCRC.
115 ADI Membership Roster, 1945, Leon Gordon Miller papers, SCRC.
116 Craig Robertson, 'Will Miss File Misfile? The Filing Cabinet, Automatic Memory and Gender', in Sarah Sharma and Rianka Singh, *Re-Understanding Media: Feminist Extensions of Marshall McLuhan* (Durham, NC: Duke University Press, 2022), pp. 119–42.
117 Leon Gordon Miller to Executive Committee (28 October 1960), Leon Gordon Miller papers, Syracuse, NY.
118 James Holland, *Minerva at Fifty: The Jubilee History of the Society of Industrial Artists and Designers 1930 to 1980* (London: Hurtwood, 1980), p. 56.
119 The IDSA held its meetings in the Playboy Hotel on Lake Geneva, Wisconsin (10 October 1968), and in 1966, 'lunch in playboy club' for the IDSA Education Committee National meeting, Parsons School of Design (6–7 April 1966), IDSA Archive, Syracuse, NY. Members' wives were traditionally invited to these events

so that they could see 'what their husbands did all day'. Shopping trips and fashion shows were organized on some occasions. The SIA also invited member's wives in Britain. Alice Twemlow comments upon this practice at Aspen in *Sifting the Trash*, p. 91.
120 American Ceramics Association Newsletter (April 1956), IDSA Archives, SCRC.
121 Letter from Eugene Gerbreux (11 October 1954), IDSA Archive, SCRC.
122 These were: Freda Diamond, New York; Elizabeth Drake, New York; Floydia Etting, Chicago; Belle Kogan, New York; Margaret Page Seagress and Beverly Ann Willis, San Francisco; Mrs Sophie Koch-Weser, Ohio; Miss Rita Long, New York and Mrs Joan F. Shortleff, Illinois.
123 Belle Kogan to the IDI board (6 July 1964), Leon Gordon Miller papers, SCRC.
124 Darling and Whitworth (eds), *Women and the Making of Built Space in England*.
125 Jessica Kelly, 'Introduction', Special Issue, 'Behind the Scenes: Anonymity and Hidden Mechanisms in Design and Architecture', *Architecture and Culture*, 6:8 (2018). See also Jessica Kelly, *No More Giants: J. M. Richards, Modernism and the Architectural Review* (Manchester: Manchester University Press, 2022).
126 Katarina Serulus, *Design and Politics: The Public Promotion of Post-War Industrial Design in Belgium (1950–1986)* (Leuven: Leuven University Press, 2018).

Professionalization of Design 1930–1980

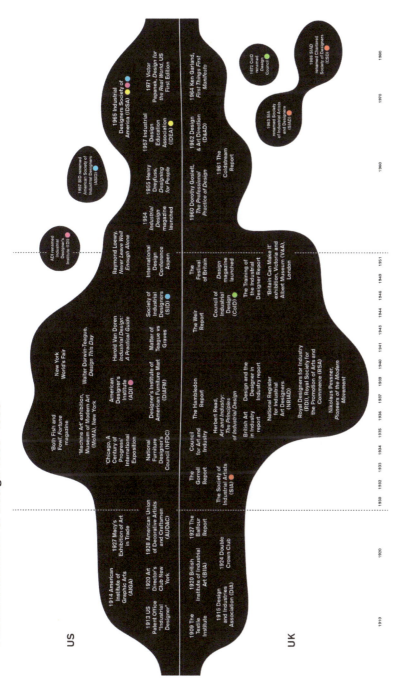

1 Timeline, 'Professionalization of Design in US and Britain, 1930-1980'. Graphic design: Katrina Wiedner. Copyright: Leah Armstrong.

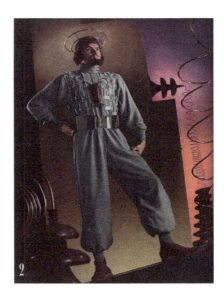

2 Gilbert Rohde models a costume from the Saks Fifth Avenue men's shop. Plexiglas for vest from Röhm and Haas, 'Fashions of the Future', *Vogue* (February 1939), p. 72. Photographer: Anton Bruehl. Image courtesy of Condé Nast publications. Copyright: Anton Bruehl, with kind permission of Anton Bruehl, Jr.

3 Advertisement, 'Olin Cellophane Speeds Package Restyling' (1956). Courtesy of Deskey Branding, Cincinnati, OH.

4 Advertisement, 'Olin Cellophane Sells the Truth', featured in *Time* Magazine (1952). Walter Landor Archive, personal papers, Box 62, Folder 11, Smithsonian National Museum of American History. Courtesy of Smithsonian Institution, Washington, DC.

5 Advertisement for Du Pont, 'Russel Wright's new plastic design goes nature one better' (c.1955). Press clippings, Russel Wright Papers, Box OS 39, Special Collections Research Center, Syracuse University Libraries.

6 Advertisement for Smirnoff Vodka, featuring Lucienne and Robin Day, *Punch* (December 1955). Courtesy of the Advertising Archives, London.

7 James Ward, cover of *Industrial Design* magazine (October 1959).

L O E W Y

question?' and I would then be stuck with wriggling out of it.

"Loewy prepared for his presentations just like a boxer preparing for a big match. He wouldn't eat; he'd be at the office hours before everyone else; he'd practice every word, every joke. When the rest of us straggled in bleary-eyed at nine o'clock, he was there, fit for a championship fight. Then, if he was nervous about his speech, he'd start off by saying charmingly, 'Please forgive my terrible French accent. I've only been in this country thirty years.' That *always* got a laugh.

"Loewy was a great manipulator—one of the best—and he learned how to manipulate the public's image of himself. He decided at the age of about twelve or thirteen to create this character called Raymond Loewy and that's what he perfected. He once told me that his mother had said that it was better to be envied than to be pitied and he made sure never to be pitied. For example, as much as he adored cars, I only saw him actually drive once or twice; he always had a chauffeur. He would meet these midwestern company presidents at the airport in his extravagantly beautiful suits and his chauffeur, and they would immediately be floored by his style.

Loewy with Betty Reese in 1946, in one of his chauffeur-driven customized cars

"Loewy made sure to orchestrate the clients' first view of his designs. If it was a car, for example, he would show it to them first from an impressive distance, from the end of a very long studio. This is because he had learned that if the clients were allowed to get close to the car in a casual way, they would start picking it apart—'What about that bumper?' 'What about this curve?' He wouldn't let them get near it until they had already experienced it as a complete and finished design. Of course, he also had a perfectly vulgar sense of what the tired housewife or the junior businessman who wanted to feel stylish would buy. He would sometimes add details to a product that he himself didn't like but that he knew would increase its market appeal.

"He was also a master manipulator of his staff. He could see dissatisfaction; he could smell disloyalty. He would take whoever was unhappy aside and gently say, 'Now, Mike, you wouldn't be thinking . . .' Or he would lean over their shoulder at the drawing board and say, 'Oh, what a beeeutiful drawing' and whoever it was would just melt.

Loewy at his stables: "He always did everything with style."

8 Betty Reese with Raymond Loewy, *Boston Herald* (17 February 1946), reprinted in Betty Reese, 'Profile: Raymond Loewy', *Industrial Design* (November/December 1986), 41. Raymond Loewy archive, Cooper-Hewitt Smithsonian Design Museum, New York.

9 Brochure: 'Industrial Design. New products: source of sales and growth' (c.1965), produced by Charles Whitney, publisher of *Industrial Design* magazine for circulation to advertisers. John Vassos Papers, Box 1, Folder 'IDSA – correspondence. General (III)', Special Collections Research Center, Syracuse University Libraries.

10 'New Space Plan: The Floating Apartment of Mr and Mrs Raymond Loewy', *Vogue* (July 1955), p. 71. Photographer: unknown. Copyright: Condé Nast Publications.

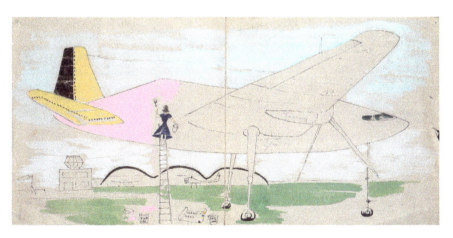

11 Gaby Schreiber, hand-painted birthday card (*c.*1950–1960). V&A Archive of Art and Design, AAD/1991/11. Courtesy of Victoria & Albert Museum. Reproduced with kind permission of Gaby Schreiber Estate.

12 Misha Black, two Christmas cards (*c.*1954–1957). Misha Black archive, Christmas card collection, V&A Archive of Art and Design, AAD/1980/3/155. Copyright: Misha Black. Reproduced with kind permission of Misha Black Estate.

4
Professional codes

Standing before a group of industrial design students at the Royal College of Art (RCA), London in May 1957, Gordon Russell, Director of the Council of Industrial Design, spoke with great conviction on the values of professionalism in design:

> In commercial art or industrial design, for the artist to become successful he must integrate himself with a team. And he has far more chance of doing that if he answers letters promptly and in good English in a legible hand or if he turns up for an interview or meeting slightly before the appointed time, if his drawings are arranged in orderly way, if he wears a neat, well-brushed suit rather than sandals and a blazer, if he is scrubbed, shaved and with well-combed hair and if he presents his case in a modest, yet authoritative way.[1]

Russell had been invited to speak to the students by School Principal Robin Darwin, to instruct them on the importance of good manners and professional behaviour, having been prompted to do so by critiques raised in the 'Student Behaviour' report by Cycill Tomrley described in the previous chapter. He was joined on the stage by SIA President Misha Black, who also advised the students on the importance of tidy dress and good manners as essential components of professional conduct. During his speech, Black raised the 'figure of the creative artist: two words which still conjure up the vision of a frustrated bohemian with a questionable private life',[2] a statement that evoked professionalization as a social project driven by Victorian moralizing ideals of self-improvement.[3] Representing the views of the CoID and the SIA, the two men's speeches conveyed a remarkable coherency on the issue of self-image, behaviour and professional conduct, as they both painted a picture of the gentleman-designer. The episode, viewed with some amusement by Darwin, confirms the paternalistic and condescending attitude of design reform, driven by the impulse to tell 'young designers how to behave and what to do'.[4]

This paternalism was inscribed in the professional codes of the main professional organizations for industrial design in both Britain and the US, which operated on an informal and formal level. In a formal sense, Codes of Conduct, issued by most professional societies and organizations as a condition of entry to their membership, set the boundaries and limitations of professional identity in a given field, issuing what sociologist Geoffrey Millerson has described as 'moral directives' and 'obligatory customs'.[5] However, professions are also governed by less formal codes, performed in the rituals, lifestyles and representations used to demarcate and give meaning to professional identity.[6] Taking both views of professional conduct into account, this chapter investigates the role played by design organizations in defining professional identity for the industrial designer. As previous chapters have discussed, the agency of the professional organization was negotiated in relation to other actors that included the government, media and business. The ability of the professional organization to impose these boundaries depended on their relationship to and with these agents of professionalization.

In a practical way, professional conduct is also self-regulated within professional societies through journals and social events at which professionals observe and imitate one another. Building on this idea, the chapter explores the capacity for design professionals in Britain and the US to self-regulate, by examining the function of professional journals and trade literature as 'forums for self-definition' and sites for professionalization in industrial design, finding that while this was a prominent feature of the profession in Britain, in the US, the capacity for this type of professionalization was tightly constrained by the dominance of the corporation on channels for communication. This chapter further examines the establishment of the Code of Conduct by the SIA, SID and ADI, reflecting on how this document defined the designer's relationship to client, business, other professionals and the public. It shows how both professional cultures were governed according to masculine codes of professional behaviour; defined in Britain as a 'Gentleman's Code' and in the US as a 'Boy Scout' attitude. It then turns to the specific regulation against advertising, finding that this aspect of the Code of Conduct was the most contentious and the most loosely interpreted by designers in both Britain – and particularly – the US. The chapter explores the agency of the individual designer to navigate this heavily codified space, as they searched for 'acceptable forms of advertising', using Christmas cards, menu designs, books, advertorial features and editorial publicity to promote their work. The final section looks at the 'exportation' of British and US Codes of Conduct through the establishment of the International Council for Societies of Industrial Design (ICSID), where British professional etiquette and US definitions of industrial design were imposed upon other national organizations, exposing the socially expansionist and universalizing objectives at the root of the professionalization by British and US design organizations.

The *SIA Journal*

As graphic design historian E. M. Thomson states in her history of the profession in the US, professional journals 'function as professional communication networks, defining professions to themselves and to others' and can be read as 'mechanisms of professional self-realisation'.[7] In Britain, the *SIA Journal* became an important testing ground within the SIA membership. Although no circulation figures are available for the journal, the parochial nature of its content suggests that it was intended for distribution exclusively within the Society itself. Frequent revisions to its content, structure and art direction give some insight into the changing self-image of the Society and the professional identities of its members. The first issue, published in February 1948, stated that it should be a publication 'by and for designers', and the importance of debate and discussion as tools through which professional identity was shaped. The *Journal* also offered an insight into the overlapping professions of advertising and design, as it functioned as a space for members to display their work, including, for example, an advertisement for Crawford's advertising agency by designer and art director Ashley Havinden, Fellow of the SIA (see Figure 4.1).

The advertisement presents the characteristic feature of the 'eye' as a signifier of professional expertise and omniscience in the advertising and design professions.[8]

The role of the *Journal* as a site of professionalization was enlivened and enriched under the editorship of designer and curator Barbara Jones, who took over between 1951 and 1953, noting in the editor's introduction of her intention to 'turn the journal away from design and to the designer'.[9] She stated, 'the hideous prospect of monotony and boredom thus opened out can only be averted by your action, your thoughtful letters on momentous themes, our stern (but never vitriolic) denunciation of abuses, your reasoned arguments, your entertaining nonsenses'.[10] A special issue in 1952, entitled 'The Designer', included articles by illustrators including Lynton Lamb on 'The Designer and his Car', Laurence Scarfe on 'Holiday Resorts', James Boswell on 'Party Conversation' and the interior decoration of the typical designer's home by architect and interior designer Hugh Casson; topics that give an insight into the aspirational, gentlemanly cultures of professionalism in British design at this time. As Lamb wrote:

> Have I unwittingly hit on something of importance to struggling designers? Would it be indiscreet to wonder whether that old Rolls helped a tiny bit towards a recent Knighthood or whether the undisputed eminence of Wells Coates is entirely uninfluenced by that 1924 Lancia?[11]

The intersection of class and masculinity with the professional identity of the industrial designer is clear here. These witty, inward-looking and highly self-referential articles give a colourful account of the lifestyle

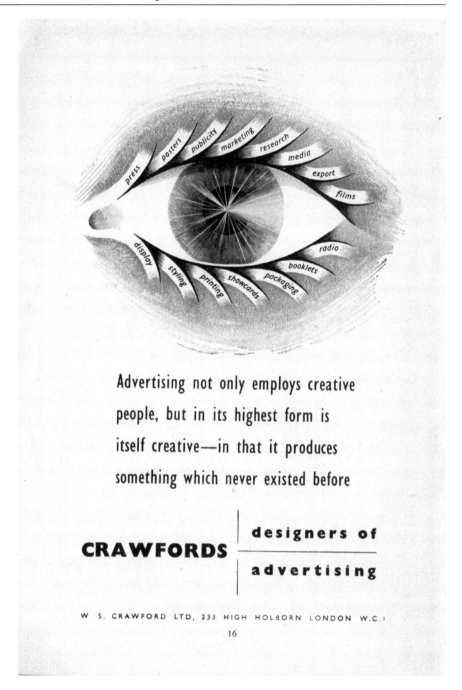

4.1 Advertisement for W. S. Crawford Advertising Ltd, London by Ashley Havinden, *SIA Journal* (February 1950), p. 18. Reproduced with kind permission of Ashley Havinden Estate and WPP plc

Professional codes 131

PARTY CONVERSATION

by James Boswell

The tailor nearly burst an artery when I asked for lapels on the waistcoat but I told him design is my business.   I'm having mine made with sleeves in.

* * *

To think I took the trouble to introduce him to my barber and then he puts in a quote a third below mine.

* * *

Where do you get such superb ties?
This thing? Oh it's from a little place I know near the Duomo in Milan.

* * *

Old chap, coloured shirts are out the day your fees get into three figures.

* * *

Well no, I don't smoke it but it's rather a pretty thing, don't you think.

* * *

Off some peg or other in Charing Cross Road I'd say.

* * *

Phoebe says it's not really mink but editors never see anything but the hat.
Well, they certainly can't ever look at the drawings she sells them.

* * *

One of the typists knits them for me.

7

4.2 James Boswell, 'Party Conversation', *SIA Journal* (August 1952), p. 7. CSD Archive. Copyright: James Boswell, with permission from James Boswell Estate, Tate Britain characteristics of 'being a designer' and provide evidence through which to substantiate claims that professional identity is defined as much through home interiors, fashion and the cars they drive, as through disciplinary knowledge.[12]

Boswell's piece on 'Party Conversation' made direct links between dress, image and professional status, ironically claiming, 'Old chap, coloured shirts are out the day your fees get into double figures.' Boswell expresses a sense of anxiety about the importance of image in securing the designer–client relationship, with both haircuts and beards being cited, echoing the concerns of Misha Black, Cycill Tomrley and Dorothy Goslett described in the previous chapter. Indeed, anxieties and concerns about professional identity and self-image frequently crossed over with what could be read as a 'crisis of masculinity' in the *Journal*. In 1953, graphic designers and illustrators exchanged letters on the subject of hats, moustaches, beards and suits. Robin Jacques, an illustrator, wrote of the 'passing sartorial expression on the part of the British male', which he attributed to the preference for professional anonymity and standardization. 'We peer dimly in wonder and awe through our National Health glasses at the days when men were unafraid to be taken for what they are.'[13] This tension between the 'colourful' and more decorative identity of the illustrator and the sober, suit-wearing industrial designer played out in debates about the cover designs, which members of the Society working in illustration sought to decorate. Misha Black, industrial design consultant and founding member of the Society, wrote to plead for a 'reticent anonymity in the Journal's future cover designs', advising it sensible to 'close digestive ramblings in a more poker-faced exterior', signalling a tension between the graphic and industrial design contingencies of the SIA membership.[14]

This preference for anonymity and businesslike behaviour coincided with the establishment of the General Consultant Designers Group, described in Chapter 2, and the formation of a Public Relations Committee, which made the designer's personal appearance its top priority. As chair of the committee Nicholas Bentley stated in the *SIA Journal* in 1954:

> Prompt replies to letters and the keeping of copies, the rendering and proper keeping of accounts and punctuality in delivering work and in appointments are not merely elementary conveniences in business, they are matters of common sense. They are also an important part of public relations. Upon their performance or neglect the business man's opinion of the artist will inevitably be coloured. The first recommendation of the Public Relations Committee is, therefore, that members should do their best to see that such opinions are of the right colour.[15]

In February 1955, student member Len Deighton contributed a cover for the *Journal* to respond to Bentley's statement, portraying a bearded man in a bowler hat standing in front of an image of himself on canvas.

The depiction articulated the tension between the image of the artist and the professional that had divided identities in architecture and design since the nineteenth century.[16] Deighton, a recent RCA graduate (and later successful spy-thriller writer), may have been one of the 'delinquent' and bearded graduates referred to in Cycill Tomrley's *Student Behaviour* Report. Inside the *Journal*, he explained that the beard shows that 'the man is an artist and the bowler hat shows that he is trying to live up to the last paragraph of Nicholas Bentley's article'; a playful example of the critical role student and graduate designers now occupied within the profession.[17]

Industrial design

Given its fixation on public image and publicity, it is something of a contradiction to note that neither of the professional organizations for design in the US, the IDI and SID, published a journal. Members of the SID frequently expressed frustration at this fact, referring to the need for an independent, professional journal, like the British *Art and Industry* and *Design* magazines in Britain, or *FORM*, the design magazine subsidized by industry in Sweden, and the newly merged IDSA made the formation of a journal a consistent goal between 1965 and 1969.[18] During this time, *Industrial Design* magazine had come to dominate industrial design as the leading source of news, reviews and commentary in the field. As Alice Twemlow writes, the magazine began as a single column in *Interiors* magazine and was developed, under the advice of George Nelson, into a publication 'concerned with product planning, design, development and marketing'. In 1954, the column editors Jane Thompson and Deborah Allen became the new magazine's editors, with Nelson as 'editorial contributor and advisor'.[19] In the absence of any state-funded publication or professional journal, the magazine took on the status of a 'mouthpiece' for the profession, even though its content was significantly shaped by advertising and commercial imperatives and its circulation and readership extended far beyond the limits of professional organizations, unlike the *SIA Journal*.[20] This dominance and influence provide some context for the IDI/SID's inability, or reluctance, to pursue a separate professional journal that might have to compete with the magazine for an audience. Instead, by 1965, upon their merger to form the IDSA, Board members tried to find a way to establish a professional journal through Whitney's patronage.

A meeting between Whitney, Henry Dreyfuss and John Vassos in April 1965 reveals the uneven power dynamics between the design organization and *Industrial Design* magazine. The men exchanged critiques of each other's work, with Dreyfuss suggesting that *ID* was 'too avant-garde for public taste'. Whitney responded by conveying his disappointment with the industrial design profession, which he had hoped would open more doors to advertising revenue. He put it to Dreyfuss that 'only one designer had ever been of any help ... by going to US Steel and suggesting they advertise in *ID*', and he thought more designers should do this.[21] Vassos turned the conversation to the central purpose of their meeting – the establishment of a professional journal for the IDSA. As Vassos explained, this would include 'few illustrations' and instead offer 'serious, learned text' written by and for the professional design community. In a less than enthusiastic response, Whitney suggested that the *ID* could include some of these articles, but Vassos's suggestion that the IDSA occupy a 'tipped-in' section of the magazine was 'not accepted too well'.[22] It was clear that the need to establish a critical, independent space for the profession could not easily be reconciled with economic advertising imperatives. This was confirmed in a letter from Whitney to Dreyfuss immediately following the meeting, in which he included a brochure, 'New Products: Source of Sales and Growth', a 'condensation of a slide presentation which we recently created, which we are showing advertisers from coast to coast' (see Colour Plate 9).

The brochure claimed that 'an estimated 75% of industry's increase in sales volume during the next three years will come from new products', making clear the relationship between the 'new profession' and its commitment to planned obsolescence.[23] As the image on the cover of the brochure conveys, the 'new profession' of industrial design was principally of interest to Whitney as a 'product' through which to sell advertising space. In 1967, George Nelson was appointed by the IDSA to adopt a new approach to the establishment of an independent professional journal.[24] He approached Arthur Drexler at MoMA to take the position of full-time editor, although this seems to have stalled on financial issues. Again, a year later, Arthur Pulos wrote to John Vassos urging the need for a 'journal to establish intellectual and philosophical substance to industrial design'; a project that was never realized.[25] The IDSA's struggle to establish a professional forum for its membership reflects the limited scope for self-critique within the profession, a point Dreyfuss also raised at the meeting with Whitney.[26]

The SIA Code of Conduct

While, as Chapter 1 outlined, the professionalization of industrial design was well under way in Britain before the war, through publications, lectures, meetings and social events, it was not until after the war, in 1945, that the Society formally published its Code of Conduct. The Code was published in brochure format, to be distributed by members to clients, as was common practice in the architectural profession. The clear structure, neat presentation and precise wording of the document suggests that detailed preparation went into its production, although evidence of this (in wartime conditions) does not survive in the Society's archive. It is likely to have been modelled on the RIBA Code and that of the Institute of Electrical Engineers, two organizations to which, in later reflections, founder Milner Gray frequently made reference. The fact that many founding members of the Society were also members and Fellows of Professional Societies in Engineering and Architecture ensured the direct transmission of these professional ideals to design. There were nine clauses in total, putting in black and white the limits and boundaries of professionalism in design for the first time in Britain. Forbidden behaviour included 'supplanting another member's work', 'breaking client–designer confidentiality', 'plagiarising', 'receiving favours or discounts in exchange for work', 'working without a fee', 'advertising' and finally, 'taking part in competitions not in accordance with the SIA's published regulations on design competitions'.[27] As a constituent of professionalism, the public featured very little in the SIA Code, which was directed towards clients, business and government. This is understandable, given the Society's origins as a trade union-like institution, with its main remit arising from the need to protect the professional practice and status of the designer. Moreover, the SIA coexisted alongside the government-sponsored Council of Industrial Design, which had a public remit.

The prohibition of advertising and publicity has been a central tenet of professionalism, established by the older professions of law, architecture, engineering and medicine. In 1945, the Architectural Association disciplinary committee stated:

> Every profession has practices which it bars. Among the commonest of these are advertising, poaching and undercutting. These activities are considered in the business world to be laudable examples of facie contrary to public policy and have always been considered offensive professionally. If a man joins a profession in which the use of trade weapons is barred and then proceeds to employ them, he is taking an unfair advantage over his Fellows.[28]

The SIA adherence to this pure professional ideal was in some ways at odds with the commercial culture of the profession, which had grown alongside, and in many cases directly within the advertising profession. Some of the Society's founding members, for instance, including Ashley Havinden, were Art Directors who regularly worked for advertising agencies on a freelance basis (see Figure 1.1). Defying its own status as a 'new profession', the Code of Conduct therefore formally aligned industrial design in Britain with the older professions of law and architecture, setting a high standard of professionalism for its members to adhere to. Accordingly, the SIA Code of Conduct was underpinned by a strong moral, rather than practical ethos. Reflecting on this later in the Society's history, Milner Gray stated that the Society's early interpretation of professionalism was 'largely governed by the descriptive and prescriptive phrase, "gentlemen will not and others must not", in an age when "gentlemen preferred blondes" and this sort of statement conjured up no archaic overtones'.[29]

The SIA operated with a highly paternalistic attitude to its membership, seeking to guide their behaviour as gentlemen professionals. Drawing on similar publications within the architectural and engineering professions, the Society published wider guidance on professional behaviour and conduct, also aimed at clients, including 'Working with your designer', which specified the working conditions and environment the industrial designer needed and should expect from a client.[30] These patrician attitudes were further elaborated upon in Dorothy Goslett's *The Professional Practice for Design*, discussed in the previous chapter. Authoritative and dictatorial in tone, it repeated SIAD policy verbatim:

> High standards of behaviour are axiomatic and to ensure that the affixes of membership are a meaningful symbol, the Society lays down a Code of Professional Conduct. Observance of this Code is a condition of membership and those who infringe it may be reprimanded, suspended or expelled. In the widest sense, behaviour includes a concern for the continuing development of knowledge and skills which future generations of industrial designers must acquire.[31]

As Goslett makes clear here, the Code inscribed a set of behaviours and attitudes that formed a blueprint on which designers should mould their professional identities. In reality, as would become increasingly clear over time, the extent to which the Code was followed 'to the letter' or 'in spirit' was a matter for debate. Engineering design consultant Jack Howe, President of the SIAD 1964–1965, insisted that it should be followed in both: 'professional integrity must be above question'.[32] For Howe, however, the Code only formalized and laid out in written form the 'professional ideal' that pre-existed any organization. It was, for him, a question of ethics: 'A thoroughly professional man does not require a Code of Conduct because it never occurs to him to do any of the things that are forbidden.'[33] Adherence to this view depended, once again, on personal commitment to and investment in the 'professional ideal'.

Professional codes in the US

At the meeting at MoMA to discuss the profession of industrial design chaired by Edgar Kaufmann, Jr in 1946 (described in Chapter 1), Raymond Loewy and Walter Dorwin Teague acted out a short improvised skit in which they dramatized the 'designer–client relationship'. In effect, the performance presented the SID ethical code 'in action'. With Loewy in the character of designer and Teague the client, the performance enacted tenets of professionalism in the SID's newly instituted Code of Conduct. The dialogue between the two men brought into tension the professional ideals of the Code of Conduct and the commercial demands of industry in which designers did their business. Teague, the client, probes the boundaries of ethical practice in design, asking Loewy to produce sketches in advance of the contractual agreement and asking him to start work on a trial basis. Loewy retorts: 'In regard to your asking us to make some sketches and then buying them if you feel that they are good or that they represent possibilities, I am sorry to say that we can't do that. It is against the ethics of the profession. One of the most important items in the code of ethics that I have mentioned is that one designer cannot work with several manufacturers in the same field.' Loewy explains, 'They come to agree on a royalty basis – a few thousand dollars a year – so that they get paid according to the success of the design. As designers we would welcome such a financial arrangement, which has been extremely successful in the past. It is one of the most successful arrangements we have in our organization. We like that sort of deal.'[34]

Teague confidently remarked to the audience, 'that was entirely unrehearsed', as he and Loewy seemed to enjoy demonstrating to their peers their self-appointed status as exemplars of ethical conduct.[35] This exchange further confirms the special function of the Code of Conduct for the SID, as a protective document that safeguarded the status of the Consultant Designer, separate and independent to the corporation. At one point in the performance, Teague asks Loewy if he feels 'he could get along with our technical staff', to which Loewy explains that he is not interested in taking credit and thereby poses no threat to the organization's internal hierarchies ... 'I'm not interested in getting a job in your company; I have my own, which is doing reasonably well', a statement that probably elicited knowing smiles in the audience, who surely knew of Loewy's well-publicised commercial successes.[36]

Speaking at the same meeting in 1946, John Vassos, founder of the ADI, was keen to stress the professional status and authority of his institution, of older standing than the recently formed SID. During the meeting, he referred to their Code of Ethics as a signature feature of its professionalized status. Vassos may have been referring to the Institute's Standards of Professional Practice, which was included in its early brochures for membership. This document set out the basic parameters for professionalism in a relatively open and loose sense, obligating members to be in 'good standing', a qualification that might have been interpreted as relating to the payment of membership fees. The introductory text referenced the designer's 'moral and cultural responsibilities to the public' and client in relatively open terms: 'The relation between the Client and the Industrial Designer shall always be predicated on good faith with the sole purpose of serving the interests of his client to the utmost of his professional ability.'[37] Like the SID and SIA, the ADI also prohibited members' participation in competitions not officially endorsed by the Society and discouraged free-pitching and speculative work. An ethics committee was established to monitor grievances and the professional behaviour of the membership. Significantly, however, the ADI did not explicitly prohibit advertising or publicity, contrary to the standards of professionalism laid down in architecture and engineering. This fits with the Society's character, which was generally less prescriptive and elitist than the SID, a fact that contributed to a more geographically, disciplinary and gender-diverse membership.

The SID drafted a Code of Conduct upon its institution in 1944 and this was circulated to members as a condition of membership, written into the by-laws of the Society. Entitled 'The Industrial Designer's Code', the opening paragraph read:

> The Industrial Designer renders professional service to his clients and through them to the public. Thus he carries a heavy responsibility for good design and good taste in the man-made world in which we live. To fulfil this responsibility and further extend its influence requires of each industrial designer in the profession high standards and ideals of practice and commission.[38]

The Code was structured in three parts, 'The Industrial Designer and His Client', 'The Industrial Designer and His Colleagues' and 'The Industrial Designer and the Public', directly addressing the three main audiences of professional identity. The Code's main concern with the 'public' related to issues of publicity and self-promotion. Directly referencing the publicity surrounding the New York World's Fair, which had centred on the image of the industrial designer as a 'forecaster', it stated:

> the industrial designer shall refrain from making any forecasts or prophecies, for advertising use, concerning future designs or future projects, which are not based on thorough research and analysis. Extravagant and irresponsible predictions of future developments discredit the individual designer and the profession.[39]

This statement appeared to be directly referencing the elaborate and wild speculations of 'Fashions of the Future' described in Chapter 1. Indeed, as *Vogue* editor Edna Wooman-Chase pointed out in her autobiography, the idea for the feature emanated from managing editor Jessica Daves and not the designers themselves.[40] The Code further restricted the designer's participation in competitions or exhibitions without prior consent of the Society, and stated, '[the] designer shall be scrupulous to avoid claiming credit not due to him or claiming more responsibility for a particular design than is rightfully his', an area of the Code that would regularly bring its members into conflict, eventually contributing to Loewy's resignation in 1964.

In 1948, the SID set out to revise the Code so that it would fit the purpose of client-facing audience even more precisely. In the course of the revisions, the Committee's entire focus turned to the 'Code of Obligation' section, which described the professional relationship between the designer and the client. In a letter to Philip McConnell in January 1948, Dave Chapman stated that the Society had 'concerned ourselves too greatly with what we expect from the public and industry in our meetings', arguing that the Code of Obligation should function as 'the logical subjective half of a contractual agreement with a client'. Another member, Dave Dailey, suggested this might be even more directly titled 'Code of Obligation to Clients', since the new code had been written 'entirely for client consumption'.[41] As this exchange makes clear, the SID was moving its focus even further from social concerns. Chapman and Dailey pushed for the inclusion of the Code of Ethics content in the Code of Obligation, stating, 'if we construe our Code of Ethics as a pattern of behaviour between ourselves, we then might consider the Code of Obligation as our behaviour towards Industry'.[42] Newly printed copies were designed to be distributed by designers and given to clients, as a solid basis for establishing a professional relationship. In 1948, Chapman, Peter Müller Munk and Brooks Stevens requested 50–100 copies each, but there is no evidence beyond this of how extensively this Code was circulated or ready by the clients it sought so directly to address.[43] This orientation of the direction of obligation from public to industry reflected the SID's consistently pragmatic, transactional approach to professionalism.

Where the SIA Code reflected the Society's values of gentlemanliness, the SID Code was described by one member, in jest, as a 'Boy Scout' code, in language that underlines the masculine cultures of both.[44] Its regulations aimed to limit competition between members of the Society, soliciting work from a company that already employed a member of the SID, and should not 'falsely nor maliciously injure, directly or indirectly, the profession, reputation, prospects or business of a colleague'. Like the SIA also, it was guided by precedent in the 'older profession' of architecture. In 1954, secretary Sally Swing ordered 25 copies of the American Institute of Architects (AIA) public relations handbook, which issued guidance to architects on how to promote their work in a professional way.[45] Swing wrote to the AIA office to state that there had been 'impressive demand' for this document by SID members, particularly from those leading their own offices as design consultants. The SID document, 'What and Why Is an Industrial Designer?' (1946), was also modelled on the AIA pamphlet, 'Facts about your architect and his work'. This document stated that 'industrial design is a profession fully as complex as architecture and engineering' and bears considerable resemblance in its function and format to SIA publications, including 'Working with your Designer'.[46]

The SID produced further guidance documents to prescribe in great detail the terms and conditions of the designer–client relationship. This included a detailed series of 'contract forms' prepared by Russel Wright to be circulated by designers to their clients on the basis of consultation (straight time; retainer; flat fee; royalty) in building construction and interior design.[47] In the case of contract for consultation, Wright advised that designers begin the relationship with a letter of agreement, rather than a formal contract, since this nature of work normally anticipates a 'longer relationship', and would benefit from a more personal approach. A further publication, 'How

To Use Your Industrial Designer', directly, through a series of questions and answers, addressed the practical concerns of industry but discouraged employers from engaging designers on the basis of 'free-pitching' or hiring more than one designer for the same job. This pamphlet was still in circulation in 1963, and was said to have been mostly used by designers for 'promotional purposes', indicating once more the value of professionalism as a promotional tool in the US industrial design profession.[48]

Acceptable advertising

Almost as soon as it was instituted, it was clear, in both countries, that the prohibition against advertising would be the most problematic feature of the Code of Conduct for the practising designer. Working within the narrow confines of what was acceptable, designers had to be inventive in promoting their work. In Britain, menus and Christmas cards, circulated between designers to each other and to clients, became a prolific site of self-promotion. The SIA Code explicitly allowed for the sending of 'change of address cards' and 'Christmas cards', both of which were identified as 'acceptable forms of advertising' in Dorothy Goslett's 1963 career-guide manual.[49] In 1956, Cycill Tomrley, manager and 'gatekeeper' of the Record of Designers at the Council of Industrial Design (CoID), wrote to the letters page of the *SIA Journal*, 'Thank you for the Christmas greetings. Those done by designers were even more delectable than usual', a subtle hint at the promotional value invested in the card as professional artefact.[50] The archives of industrial designers often contain evidence of this in the form of the 'Christmas card list', carefully maintained and managed as a roster of professional networks.[51] Misha Black's archive at the V&A Archive of Art and Design in particular contains a considerable number of Christmas cards, which he designed throughout the 1950s and 1960s. As two images from 1954 and 1957 show, Black often incorporated a self-portrait in his Christmas card design and appears to have spent considerable energy in designing them, as suggested in the caption on one card, which reads, 'Gnawed by anxieties – worrying how to wish you a Merry Christmas' (see Colour Plate 12).

Similarly, menu designs served a promotional function for designers in Britain. This can be most clearly seen through the archive of the Faculty of Royal Designers for Industry at the Royal Society for the Promotion of Arts and Commerce, London. The Double Crown Club had long held the tradition of inviting members to design menus for its monthly luncheons and the Faculty incorporated this tradition in 1961, when Ashley Havinden was invited to design the menu for the Faculty's twenty-fifth anniversary dinner, a tradition that survives until this day.

As early as 1946, only two years after the establishment of the SID in the US, members participated in a panel discussion entitled 'When is advertising not advertising?' The Society's Code of Obligation decreed that '(the designer) shall not approve nor permit publicity to be released which might injure the dignity and standing of the profession as a whole'.[52] In his opening remarks, Harold Van Doren summarised the problem that faced 'every freelance designer' in bringing his services to the attention of potential freelance clients. As acceptable alternatives to advertising, Van Doren noted that the SID allowed the following behaviour: 'Personal contact with industrial executives; Satisfied clients talking with potential clients; Newspaper or magazine publicity; Publicity resulting from winning an award; Articles written by the designer or members of his staff; Mention of the designer in a clients' advertising; Publication of a book.'[53] This list of activities, some of which would later be refuted as beyond the limits of professionalism, provides a clear indication of the significance of social activities and self-promotion in design work. Voicing the concerns of some members who were seeking a more relaxed attitude to advertising, Van Doren expressed his belief that 'some professions have changed their attitude about advertising', citing banking as an example, and wondered if it might be a good idea for designers.[54] Egmont Arens argued 'publicity is very important for client attention ... Well-handled publicity for one industrial designer benefits the whole field.'[55]

During the discussion, Walter Dorwin Teague interjected to relate his experience of the New York court case in which he sought to overturn the taxation of the designer as a non-professional. During the case, he stated, the question of whether he used paid advertising came up repeatedly and the fact that he did not advertise was considered 'a favourable factor in the final decision that industrial design is a profession rather than a business'. Rather than advertising, he explained, legal experts considered his book and magazine articles as evidence of the professional nature of his field. Thus, while designers should take a 'strong stand against undignified publicity', it was important to 'maintain a designer's media visibility, while strictly forbidding the use of advertising'.[56] For SID members, it was not a question of whether the designer should advertise, but a question of what kind of publicity was appropriate and effective. In July 1962, Dreyfuss wrote to Arthur Becvar, 'as you know, I am thoroughly against any Society director interfering with the operation of an individual's office ... It is virtually impossible to control the press. Often a general article on a designer's work will include many products already publicized and so clearance is unnecessary.'[57] This flexible approach to professionalism characterized attitudes to professional conduct and behaviour within the SID, in contrast with the more rigid interpretation followed by designers in Britain.

Breaking the Code

In 1963, designer Terence Conran, who had been appointed a Fellow in 1951, was dismissed from the SIAD for advertising and self-promotion. Conran remembered the episode vividly in an interview in 2012:

> One day I came into the SIA office and the secretary held the document in front of me and said, 'What is this?' And I said, 'it's an advertisement for my work', and he said, 'Well you're not allowed to do that. Don't you know that you were advertising to a company where they already employ an SIA member?'[58]

Conran was personally hurt by this public dismissal, especially as he had only recently commented in the Society's journal on the need to tighten up the membership entrance requirements.[59] Nevertheless, his public ousting had no demonstrable impact on his success as a designer. Architect and designer Jack Howe, who was President of the SIA at this time, held particularly strong views on the values of professionalism. He wrote to David Harris in 1965, 'I don't think the Design Council or the SIAD are doing very much to help (the profession). They are busy courting the Conrans and the rich groups with little regard for the quality of work produced.'[60] In 1965, the Society published a letter, Clause 13a of the Code of Conduct, distributed to all members, as a draft example of the 'acceptable means of advertising one's services'. Quaintly formal, it articulated the manners of the gentleman designer:

> Dear Sir, May I bring to your notice the design services I can offer in the following fields. You will appreciate that this letter, which conforms to the Code of Professional Conduct of the SIAD, is written on the assumption that you do not already retain a graphic industrial designer in the above fields. Should this assumption be incorrect, I would not wish to pursue this matter.[61]

There are no records in the archive indicating how often this letter was used by members. In a letter to architect and Fellow of the SIA Erno Goldfinger, textile designer John Tandy wrote, 'I believe nearly half of the total membership of the Society feel that Clause 13a (advertising) is unrealistic and undesirable.'[62] The question of advertising precipitated a professional crisis within the SIA, but it also reflected broader social and cultural change in Britain, as the values of professionalism and working culture were increasingly driven by the business-oriented dynamics of free-market capitalism, over the gentlemanly pace of design reform. Stuart Rose summarised this in his Presidential address of 1964:

> Can we be sure that we are furthering the aims of industrial design in general and of our members in particular when on the one hand we strive to persuade industry that design is an essential part of industrial production and on the other hand we try to emulate the detached independence of the older professions? For whereas our professional inclination is to wait behind the brass plate of our professional door, our concern for higher standards of design and our belief in our ability to create them inclines to force us into the market place, where industry does its shopping.[63]

By 1966, the disciplinary powers of the Code of Conduct had been strengthened, but also generalized, stating that 'Council may reprimand or expel any member who conducts himself in any manner which discredits his profession', so that the question of advertising and self-promotion was now to be governed by common-sense pragmatism, rather than formally encoded professional ethics.

The regulation against advertising was also the SID's most divisive instrument. Attitudes to advertising, self-promotion and professionalism were divided according to the identity of the corporate (in-house) designer and the freelance consultant. This was plainly expressed in a heated exchange of letters between Don McFarland, in-house designer at General Electric, and Walter Dorwin Teague, head of his own consultancy office. The attitudes of these designers came into conflict during a professional committee meeting discussion on the recent application to membership by the well-known designer Gordon Lippincott of Lippincott–Margulies, a firm 'styled after an advertising agency and very promotionally minded'.[64] McFarland, who sponsored Lippincott's application, accused the Society of hypocrisy, given that 'our code of ethics has never forbade direct mail advertising … because many of our members also indulge in it and have never been censored for it'.[65] Teague wrote to Ray Spilman, 'I have great respect for Don McFarland and for his professional standards, but I realized for a long time that he is not a consulting designer and does not share our interest in building up his professional stature.'[66] As he explained to Spilman, the consulting designer must have complete confidence of the industry in his professional integrity and the prohibition of advertising was essential to this integrity. By contrast, he explained:

> Don has an excellent job in which he is turning in a fine performance. I am sure he has a sizeable salary with many fringe benefits and his future is secured up to and beyond retirement age. Therefore, having the complete confidence of General Electric he does not feel the necessity which the rest of us are under to win and hold the confidence of industry in general.[67]

In the same letter, Teague admits to giving permission to US Steel to use an interview with him on the use of materials for a forthcoming magazine advertising campaign, which, he says, was 'no more objectionable than appearing in the editorial pages'.[68] 'Don probably would not approve of this, but he is not under the necessity of maintaining a reasonable flow of clients into his office.'[69] It seemed that while Teague upheld the principle of prohibiting advertising, he was prepared to be flexible when it came to his own business. As McFarland pointedly stated in a letter to Teague, 'perhaps it is characteristic of consultants to evaluate the profession from their own eyes'.[70]

Tensions between the corporate and Consultant Designer – already heated – were inflamed by the question of advertising and self-promotion. Indeed, the SID's operation of a 'double-standard' approach is easily traceable through the archive records of the SID disciplinary folder, in which members wrote to complain of Raymond Loewy's publicity campaigns in print, radio and television, but senior members of the SID are reluctant to take any action. In December 1953, Henry Dreyfuss wrote to Sally Swing, drawing attention to a recent article in *Architectural Forum* warning of the use of architects' pictures in advertising as 'something akin to bribery' and characterized them as 'flagrant transgressions against the tenets of good professional taste', a statement that must surely have been tongue-in-cheek, given the prolific appearance of Consultant Designers in public advertisements.[71] In 1957, President Jay Doblin wrote to Brooks Stevens to warn him that his recent advertisement in a local newspaper went against the Society's Code of Ethics. He told him:

> There are many different ways of making oneself known, through speeches, letters and so forth, which ASID sanctions, but in the matter of paid advertising, we have gone along with the professions of architecture, medicine and others in not permitting direct paid advertising in newspapers or other media and have worked hard to establish and maintain strict observance of this.[72]

Stevens, who had previously been disciplined by the Society for his controversial interview on 'obsolescence' in *True: The Man's Magazine*, apologised for the 'error', explaining that it was under pressure from the publication's editorial team in exchange for free editorial previously given.[73] In 1959, Don MacFarland explained the Society's position as follows: 'If a designer has used a material in a design, an ad can say so – but it cannot include an endorsing statement from the designer ... This is like a doctor saying if I was going to operate, I would use Cut-Rites' surgical tools.'[74] This statement was clearly directed at the Consultant Designer contingent of the SID membership, who participated most enthusiastically in advertising campaigns, as described in Chapter 2.

Debates about advertising within the ASID revealed the inconsistency to which this had been applied across its membership. A growing sense of division between the corporate and Consultant Designer was mounting at a time when the role of Consultant Designer, in the face of integration as a dominant model of employment, was already in decline.[75] In July 1963, Alfred Wakeman wrote to the SID grievance committee to complain about the competitive behaviour of a fellow member, who had solicited work from a company he knew employed Wakeman. Wakeman wrote, 'We designers are evidently a badly divided group with hard driving businessmen in one camp and professionalisms [sic] in the other.'[76] Don McFarland reminded the grievance committee that there was nuance in every case and pointed to the fact that the Society's attitude to professionalism might be beginning to outdate:

> Certainly we are all aware that Walter Teague's feeling that a designer should wait for the client to contact him is not realistic in this day and age ... Also I feel that some designers regard their clients as personal property, including all divisions and product lines, despite the fact that they are working only on one product.[77]

It was clear that the Consultant Designer no longer held a monopoly on professional identity in US industrial design, as alternative voices entered the fray.

The increased prominence of the General Consultant Designer in the British design profession put pressure on the prohibition against advertising in the SIA Code of Conduct. In 1969, British architect and designer David Pearson wrote an article in the journal *Design Dialogue* in which he reflected on the contradictions within the SIAD attitudes to professionalism. 'It may be', he argued, 'that the more stalwart defenders of traditional professional position – no advertising or soliciting of work, no financial interests or discounts received etc. – are thinking of consultancy.' Pearson argued that the hierarchical position of the Consultant Designer within the SIAD organizational structure had led to the setting of an unfairly high level of ethical standards with regard to advertising, self-promotion and scales of fees. He argued that the 'consultant designer is sufficiently different from the average practitioner to justify a special code of conduct. If use of the words "consultant designer" by SIAD members was restricted to those who had agreed to be separately bound by a Consultants' Code, many anomalies would disappear.' On the other hand, he said, for 'less ambitious members who only wish to do good jobs, a more relaxed code would suffice'.[78] In this sense, the double standards that had operated in the US attitudes to professionalism within the SID were also now in operation in Britain.

International codes

By the 1960s, the 'soft power' of design was mobilized through the establishment of international design organizations and this was recognized, early on, as an opportunity through which to export the professional codes of behaviour established by British and US design organizations. The ICSID was founded in London on 28 June 1957, by designers from Europe and the US, to raise the professional status of designers, to establish international standards for the profession and to improve industrial design education.[79] The Council arrived at a definition of the industrial designer as follows:

> One who is qualified by training, technical knowledge, experience and visual sensibility to determine the materials, construction, mechanisms, shape, colour, surface, finishes and decoration of objects which are reproduced in quantity by industrial processes. The industrial designer may, at different times, be concerned with all or only some of these aspects of an industrially reproduced object.[80]

The Council was careful to find a definition that would be as holistic and inclusive as possible, by offering an extension of this definition, to 'packaging, advertising, exhibiting, and marketing when the resolution of such problems requires visual appreciation in addition to technical knowledge and experience'. Craft was also covered.[81] In 1966, Arthur J. Pulos wrote to Ramah Larisch, Secretary at the IDSA, to complain that the ICSID definition of industrial designer was too general, drawing comparisons to architecture and medicine, where, he argued, the definition was tighter and more precise.[82] The IDI was more generally sceptical of the ICSID project, claiming that many member societies of the ICSID are 'woefully substandard'.[83] Nevertheless, US industrial designer Peter Muller-Munk 'relentlessly contended' that the ICSID's adoption of the term 'industrial design' (a term not in use in most member countries at this time) ultimately 'reflected the significant influence of the US on European design'.[84]

The institution of the ICSID opened a new chapter for the industrial designer's professional identity. According to Tania Messell, the Council's early emphasis was on the professionalization of the designer and not design.[85] Photographs of Misha Black at the ICSID conference in Venice in 1961 show the enhanced status of the designer as a diplomatic bureaucrat (see Figure 4.3), a role which Black and others clearly enjoyed playing.

This new role for the designer in international relations and cultural diplomacy presented a flattering self-image of the designer's agency and influence. Black and the other founding members of the ICSID did so in search of an international standard for the designer as a kind of universalizing ideal.[86] Driven by a culture of superiority and privilege that would later come under scrutiny and critique within the ICSID itself, the SIA used the ICSID as a platform through which to export and universalize its own Code of Conduct and Articles of Association, noting with pride that the ICSID Code had been modelled on its own. In the process, they transplanted and exported the SIA's gentlemanly ideas of professional behaviour, setting a 'standard' of behaviour that derived from traditional notions of professionalism that had their origins in Western industrial capitalism. This was naturally aided by the fact that negotiations commonly took place in English, a point of great frustration and resentment by organizations in many European countries. As the first professional design society in the world (self-proclaimed), with the first Code of Conduct for the Professional Designer, Henrion stated that it was 'a natural corollary' that 'most countries' professional societies, as they began to organize themselves, based their own codes and rules and regulations on those of the SIAD'.[87] It was not the first time that the Society had shown expansionist ideas. In 1949, Milner Gray toured New Zealand in 'statesman-like fashion', to propagandize design on behalf of the British Council.[88] Writing in his diary of a visit to Buenos Aires in 1963, Misha Black described the profession as 'where Britain was in 1930, the early days of a new profession'.[89] The ICSID gave FHK Henrion and Misha Black, both émigrés, a platform on which to enact, perform and enlarge their influence as global actors in a newly

4.3 Peter Muller-Munk, Misha Black and Sigvard Bernadotte at the International Council of Societies of Industrial Design (ICSID) General Assembly, Venice (1961). Photographer: unknown. Misha Black archive, V&A Archive of Art and Design, AAD/1980/3/55.

formed international space of design diplomacy. With the opening of the European Common Market in 1957, Henrion stated, 'it is no longer possible to look at the future of design on a purely national level':

> What is done by our colleagues all over the world is brought to us by publications and our work must always relate to theirs as theirs to ours. In addition, more and more designers are already working now for clients outside the UK, so that the design profession and design services are becoming ever more international as commerce and industry spread more and more across national boundaries every year.[90]

These were bold words for a society that had always been inward-looking in its focus on national tropes of gentlemanliness.

The US professional organizations, the IDI and SID, navigated a more complex and less complimentary self-image abroad. The formation of the ICSID in particular forced US industrial designers to face difficult realities at home, particularly in relation to the divided status of its two organizations – the

SID and IDI. In being forced to share its delegation, the IDI and SID came to see the problems of its home conflict play out on an international stage. The unification of the two, in the form of IDSA (1965), was in some part a consequence of this realization. Furthermore, US delegates, including Arthur J. Pulos, reported back to members at home on the uncomfortable reputation of US industrial designers as overly commercial and capitalist in their tastes and practices, a critique politicized under the conditions of cold-war politics. As Pulos recalled of the Vienna meeting in 1965:

> The most popular subject was to lambast the Americans and no matter, whatever speaker got up, somewhere in their conversation they said evil things about the materialism of the Americans and how they were trampling all over the world and doing this and doing that and I finally had it, so I asked for the floor and I just raised hell. And I said [that] none of them would be there at all if it weren't for the Americans, and if they enjoyed beating the Americans down, it was all well and good, but please don't forget that their very freedom is something that we created. You know, a real good American red-white-and-blue speech.[91]

Pulos's emotional response here points to a frustration faced by many American designers when confronting their self-image abroad; an image that had been heavily mediated and proudly projected by its leading consultant protagonists. In this way, international design organizations opened up a new space for national organizations to see their self-image and identity in a broader context. By the end of the 1960s, they also significantly shifted the agenda of professional design organizations towards social issues and concerns, including 'world poverty, sustainable development and gender inequalities', putting the inward-focused nature of the Code of Conduct, with its dominant focus on the client relationship, in sharp perspective.[92]

Conclusion

While previous chapters have pointed to the contrast between the visibility of the US industrial designer versus the relative anonymity of their British counterpart, this chapter has suggested an inverse dynamic in relation to the operation of what sociologist Geoffrey Millerson has termed the 'inter-professional' versus the 'intra-professional' dynamics of professionalization.[93] In short, even though the US industrial designer occupied a more colourful space within the mainstream media, on closer inspection, the absence of a professional journal at any point in the profession's history points towards the tightly constrained position of the industrial designer as a 'middleman' between industry and consumer. By contrast, in Britain, where the public image of the designer in Britain was obscure and relatively anonymous, the prominent role played by reforming agents, learned societies and the SIA ensured a more self-reflective, internally critical space through which members could regulate their own 'interprofessional' behaviour. Here, the chapter highlighted an internal preoccupation with dress, hair and beards within the pages of the *SIA Journal*, as concerns about professional identity were expressed through a 'crisis of masculinity' that recurs in almost every chapter of this book.

While previous chapters identified the hyper-masculine codes that permeated the profession in media performances and exhibition characterizations, this chapter has shown that the Codes of Conduct issued by professional organizations acted to further cement these ideals, pinning the identity of the designer to the gendered ideals of the gentleman or 'Boy Scout'. However, many aspects of this code, most especially attitudes to advertising and self-promotion, were contentious within the membership and widely criticized by designers who regarded them as high-minded and not applicable to the commercially oriented practice of industrial design. Again, the question of whether design was a new profession, like advertising, or an old profession, like architecture, lay at the centre of these debates. However, the ability of these institutions to exert their influence and power was constrained by the varying scale and power of other agents – including the media, government, business and corporations. While the SIAD and IDSA issued Codes of Conduct which on surface responded to the same professional ideals – putting limits on entrance, advertising and competition – the application and enactment of these ideals differed considerably.

In particular, the relationship with the client was prioritized in the codes of conduct for organizations in both Britain and the US, as the industrial designer continued to hold an obscure relationship with the public until the 1970s. When 'designing' their Code of Conduct, designers expressed an awareness of three main audiences – the client, the public and other professions – and they were each addressed in the original documents. However, emphasis within the documents was unevenly distributed, with the relevance of the public being generally forgotten or dismissed, except in the recognition of their status as 'consumers' or 'the market'. As such, the idea of public good, a central tenet of pure professionalism, was not addressed, providing an unstable and flawed basis on which to pursue professionalization in industrial design. By the end of the 1950s, professional egos were boosted by the 'exportation' of these values and their instillation within the ICSID, assuring the cultural superiority of the *industrialized* over the *industrializing* countries. Nevertheless, as the next chapter will explore, this confidence was short-lived as international forums opened up new spaces for critique, including the newly established International Design Conference in Aspen (IDCA) forcing 'industrialized' designers to confront their new status as 'cultural workmen' in the apparatus of advanced capitalism.[94] As historian Jan Logemann so eloquently asks, 'how did the "taste-makers" become the "waste-makers" of consumer capitalism?'[95] The next chapter turns to address this question.

Notes

1. Gordon Russell, Notes for RCA Talk to Students, Gordon Russell Talks, Part D (2 May 1957), CoID Archive, UBDA.
2. Misha Black, Notes for Talk at the RCA (19 March 1957), Student Behavior Folder, Design Council Archive, UBDA.
3. Neil McKendrick, 'Gentleman Players', in Neil McKendrick and R. B. Outhwaite (eds), *Business Life and Public Policy: Essays in Honour of D. C. Coleman* (Cambridge: Cambridge University Press, 1986).
4. Ernest Race, quoted by Dorothy Goslett in her interview with Robert Wetmore (23 January 1983); see Chapter 3, p. 114.
5. Geoffrey Millerson, *The Qualifying Associations: A Study in Professionalization* (London: Routledge, 1964), p. 149.
6. See Millerson on the image dynamics of a profession, ibid., p. 1. See also Sean Nixon, *Advertising Cultures: Gender, Commerce, Creativity* (London: Sage, 2003). Leah Armstrong, 'A New Image for a New Profession: Self-Image and Representation in the Professionalization of Design in Britain, 1945–1960', *Journal of Consumer Culture*, 19:1 (2019), 104–24.
7. Ibid.
8. Catherine Moriarty, 'A Backroom Service? The Photographic Library of the Council of Industrial Design, 1945–1965', *Journal of Design History*, 13:1 (2000), 39–57.
9. Barbara Jones, Special issue, 'The Designer', *SIA Journal* (August 1952), 6–12.
10. Ibid., p. 1.
11. Lynton Lamb, 'The Designer and His Car', in ibid., p. 5.
12. David Wang and Ali O'Ilhan, 'Holding Creativity Together: A Sociological Theory of the Design Professions', *Design Issues*, 25:1 (2009), 5–21.
13. Robin Jacques, *SIA Journal*, 11 (September 1949), 3–4.
14. FHK Henrion later remarked upon this tension in *The Designer* (October 1970), 5.
15. Nicholas Bentley, 'The Public Relations Committee', *SIA Journal*, 41 (October 1954), 9.
16. See Andrew Saint, *The Image of the Architect* (London: RIBA, 1980).
17. *SIA Journal* (February 1955). See Armstrong, 'A New Image for a New Profession', p. 117.
18. SID Annual Meeting, Waldorf Astoria, New York (12–14 October 1949), IDSA Archive, SCRC.
19. As Twemlow notes, Nelson's influence was significant, as he had a design office in the same building and frequently made editorial decisions on what went in the magazine. See Alice Twemlow, *Sifting the Trash: A History of Design Criticism* (Cambridge, MA: MIT Press, 2017), p. 55.
20. Ibid.
21. Henry Dreyfuss, notes from meeting with *Industrial Design* (15 April 1965), Vassos papers, Box 1, Correspondence, SCRC.
22. Ibid.
23. Charles E. Whitney to John Vassos, with 'New Products: Source of Sales and Growth' enclosed (15 April 1965), Vassos papers, Box 1, Correspondence, SCRC.
24. George Nelson, recommendation to Board of Directors, IDSA (10 January 1967), Box 2, Board of Directors, Vassos papers, SCRC.
25. Arthur J. Pulos to John Vassos (24 September 1968), Box 2, Board of Directors, Vassos papers, SCRC.
26. Henry Dreyfuss, notes from meeting with *Industrial Design* Magazine, Vassos papers.
27. Code of Conduct (July 1945), Box 93, CSDA.
28. ArcUK discipline committee folder on Murray and Arnold Graves Avery (17 October 1962), Erno Goldfinger papers, Series 49, RIBA.

29 Milner Gray, SIAD Presidential Address, 1968, 'The Price and the Value', Box 200, CSDA.
30 'Working with your designer' (CSDA, c.1955), 4.
31 Goslett *Professional Practice of Design*, p. 3.
32 Jack Howe, Course for Design Lecturers in Professional Practice and Design Administration (25 November 1971), pp. 4–5. Jack Howe, personal archive.
33 Ibid.
34 Walter Dorwin Teague and Raymond Loewy, 'Conference on Industrial Design, A New Profession', MoMA for the SID (12 November 1946). Henry Dreyfuss Archive.
35 Ibid.
36 Ibid.
37 ADI Standards of Professional Practice, c.1940, Leon Gordon Miller papers, SCRC.
38 SID Code of Ethics (1944), IDSA Archive, IDSA_62, SCRC.
39 Ibid.
40 See Leah Armstrong, '"Fashions of the Future": Fashion, Gender and the Professionalization of Industrial Design', *Design Issues* 37:3 (Summer 2021), 11.
41 Dave Dailey to Dave Chapman (6 February 1948), IDSA Archive, SCRC.
42 Dave Chapman to Philip McConnell (25 January 1948), IDSA Archive, SCRC.
43 Letters of request from Dave Chapman, Peter Muller-Munk and Brooks Stevens, IDSA, Box 62, Folder 4, Code of Practice Requests, IDSA Archive, SCRC.
44 Dave Dailey, 'Sounds a bit like a Boy Scout code', letter to Dave Chapman (6 February 1948), IDSA_62_, IDSA Archive, SCRC.
45 Sally Swing to the AIA (17 June 1954), AIA folder, IDSA Archive, SCRC.
46 'What and Why is an Industrial Designer?' (1946), IDSA Archive, SID, IDSA_62, SCRC.
47 IDSA Archive, Box 62, 'Code of Obligation' (3 February 1948), SCRC.
48 'How to use your designer', Box 62, IDSA Archive, SCRC.
49 Goslett, *Professional Practice of Design*, p. 35.
50 Cycill Tomrley, *SIA Journal* (February 1956), p. 3.
51 Susan Wright, daughter of Jack Howe, SIAD President, said during a meeting with the author (4 June 2011) that her father spent a lot of time updating his Christmas card list and ensuring its accuracy.
52 Code of Obligation, IDSA _64_2 IDSA Archive, SCRC.
53 Harold Van Doren, SID Meeting (17 December 1945), IDSA Archive, SCRC.
54 Ibid.
55 Egmont Arens, SID Meeting (17 December 1945).
56 Walter Dorwin Teague, SID Meeting (17 December 1945).
57 Henry Dreyfuss to Arthur BecVar (9 July 1962), IDSA Archive, SCRC.
58 Terence Conran, interview with the author (14 July 2011).
59 Conran wrote to advise the SIA to 'make the entry of members to the Society very much more strict … I feel that the present entry to the SIA is much too easy and the reputation of the Society suffers as a consequence', *SIA Journal*, 98 (April 1961).
60 Jack Howe to David Harris (29 October 1965), Jack Howe, personal archive.
61 Code of Conduct for the Professional Designer (May 1965), Box 200, CSDA.
62 John Tandy to Erno Goldfinger (14 January 1966), Erno Goldfinger Archive, 2 Willow Road, c/o National Trust, UK.
63 Stuart Rose, Presidential Address (1964), Box 101, CSDA.
64 Don McFarland to Walter Dorwin Teague (10 May 1960), IDSA 62_9_005 33/38, SCRC.
65 Ibid.
66 Walter Dorwin Teague to Ray Spilman (4 May 1960), IDSA 62_9_005 33/38, SCRC.
67 Ibid.

68 This is probably the designer Charles Whitney refers to as helping *Industrial Design* with advertising.
69 Walter Dorwin Teague to Don McFarland (4 May 1960), SCRC.
70 Don McFarland to Walter Dorwin Teague (10 May 1960), SCRC.
71 Henry Dreyfuss to Sally Swing (11 December 1953), IDSA_62_7, IDSA Archive, SCRC.
72 Jay Doblin to Brooks Stevens (9 February 1957), IDSA, Ethics folder, 62_8, IDSA Archive, SCRC.
73 Brooks Stevens to Jay Doblin (11 February 1957), IDSA, SCRC.
74 Don McFarland to Walter Dorwin Teague (10 May 1960), IDSA 62_9_005 33/38, SCRC.
75 A professional evaluation committee within the ASID noted this decline in 1962, stating that while design departments were growing, there was less growth for the design consultant. ASID Professional Evaluation Committee report, 1962–1963, p. 2, Raymond Spilman papers, 'Talks 1961–2', SCRC.
76 Alfred W. Wakeman to the IDSA Board (13 July 1963), IDSA Archive, Box 62, Folder 14, SCRC.
77 Don McFarland to Walter Dorwin Teague (10 May 1960), IDSA Archive, SCRC.
78 David Pearson, 'Are designers professional?', *Design Dialogue* (Spring 1969), 15–16. David Pearson, personal archive.
79 ICSID Constitution adopted at the first General Assembly, Stockholm, Sweden (1961), ICSID Archive, UBDA.
80 Ibid.
81 'The designer for craft-based industries or trades where hand processes are used for production is deemed to be an industrial designer when the work produced to his drawings or models is of a commercial nature, is made in batches or otherwise in quantity, and is not the personal work of the artist-craftsman', ibid.
82 Arthur J. Pulos to Ramah Larisch (6 January 1966), Box 1, Correspondence, Vassos papers, SCRC.
83 Ted Clement to Leon Gordon Miller (9 March 1962), Leon Gordon Miller papers, Box 1, Correspondence, SCRC.
84 Tania Messell, 'Constructing a United Nations of Design: ICSID and the Professionalization of Design on the World Stage, 1957–1980' (PhD thesis, University of Brighton, UK, 2014), p. 67.
85 Ibid.
86 Ibid.
87 FHK Henrion, '"International Relations", SIAD: The First Forty Years', *The Designer* (October 1970), 5–6.
88 Christopher Thompson, 'Modernizing for Trade: Institutionalizing Design Promotion in New Zealand, 1958–1967', *Journal of Design History*, 24:3 (2011), 255.
89 Misha Black, typescript of diary visit to Buenos Aires (May 1963), Misha Black Archive, Personal Diary, V&A Archive of Art and Design, AAD/1980/3/65.
90 Henrion, *Designer* (October 1970), p. 6.
91 Pulos, interview with Robert Brown (31 July 1980–5 December 1982), quoted in Messell, 'Constructing a United Nations of Design', p. 113.
92 Jeremy Aynsley, Alison J. Clarke and Tania Messell, 'Introduction', in Aynsley, Clarke and Messell (eds), *International Design Organizations: Histories, Legacies, Values* (London: Bloomsbury, 2021).
93 Millerson, *Qualifying Associations*, pp. 148–9.
94 C. Wright Mills, 'The Man in the Middle', *Industrial Design* (November 1958), pp. 70–5.
95 Jan Logemann, *Engineered to Sell: European Émigrés and the Making of Consumer Capitalism* (Chicago: University of Chicago Press, 2019), p. 131.

5
Crisis of professionalism

> Industrial design is still an immature profession because until now it lacks that which characterises maturity: the capacity to recognise its own limits.[1]

> Industrial design is the only profession that reached myth before it reached maturity.[2]

These two statements, delivered separately by Argentinian-born industrial designer and President of the ICSID (1966–1969) Tomás Maldonado and US architect, designer and critic George Nelson, articulate the existential crisis that was mounting for industrial design internationally. Intriguingly, both also point to the enduring prevalence of the term 'maturity' in professional design discourse in Europe and the US. This generational framing, which enlarged the agency of the individual designer, is reflected in the history of the profession, as it has often been described as a generational conflict between the 'founding fathers' of the profession and the rebellious younger generation, who by the 1960s were calling the ideologies and principles of professionalism and planned obsolescence into question. In short, the failure to 'professionalize' is commonly framed as a failure to 'mature', revealing the close association between values of professionalization and progress, industrialization and technological advancement.[3] This sense of design as an 'incomplete' and 'unfinished' professional project continues to define and frame contemporary design discourse.

This chapter looks closely at the dynamics of failure in industrial design, focusing on the so-called 'crisis of professionalism' in both Britain and the US from the 1960s onwards. It explores attempts within the profession to reflect, respond and manage professional crisis as an opportunity for 'reinvention', as the 1960s saw the opening-up of new forums for critique. This includes debates about the role and value of the industrial designer at Aspen, Colorado, USA and within the professional organizations, the

SIA and IDSA. Here, the chapter focuses on the publication of two major reports on professionalism by these two organizations, both of which addressed the sense of professional crisis in design. Examining organizational responses to these critiques, the chapter finds that they were in most cases insufficient to deal with the existential crisis facing a profession forged to accelerate industrialization and consumer growth; two concepts now facing radical critique.

Aspen

Previous chapters have commented upon the relatively constrained nature of design discourse in the US and limited space for self-reflection or self-critique. Responding to this absence, the International Design Conference in Aspen (IDCA), founded by Walter Paepcke and Egbert Jacobsen of the CCA, was originally conceived in 1951 as a forum for designers and businessmen to discuss the shared interests of culture and commerce at a far remove from their everyday concerns. Board members over the decade included George Nelson, Eliot Noyes and Saul Bass, designers who did not subscribe to one particular professional design organization or creed over the course of their careers, indicating the increasingly limited significance of the professional organization in anchoring design ideas and debates. As an exception, John Vassos, founding member of the ADI, was involved in the Conference in the early years of preparation, but had hoped it would later move to the East Coast and later distanced himself from it.[4] As design historian Robert Gordon-Fogelson puts it, the IDCA differed in important ways from design organizations, 'whose meetings typically centered on the passage of resolutions related to formal training, best practices and other means of regulating the professions'.[5] The Aspen Conference was looser and more discursive, providing a space in which to generate debate and discussion pertaining to the advancement of design within the boundaries of the corporation and big business. In this sense, it invented a seemingly 'safe space' for professional critique, without disturbing the hegemonic tropes of industrial design in the US, which included the corporation and the dominant position of white, privileged men and their modernist tenets of good taste.

In the summer of 1958, sociologist C. Wright Mills delivered a paper at the Aspen Conference entitled 'The Man in the Middle'.[6] Introducing a reprint of the paper in its pages in November 1958, *Industrial Design* magazine described it as 'neither lullaby nor mock attack, it is a hard analysis of the designer in our society'.[7] Wright's paper contributed to an 'anxious form of critique'[8] that had been building on both sides of the Atlantic, which addressed the designer's culpability in the crimes of consumer capitalism. As Mills put it, 'Since the end of the Second World War, the new economy has flowered like a noxious weed. In this phase of capitalism, the distributor becomes ascendant over both the consumer and the producer.'[9] For Wright, the designer, a 'cultural workman', was the product of an 'overdeveloped society with its ethos of advertisement' and had contributed to the fetishization of the mass sales of goods through a blind commitment to obsolescence, joining advertising, public relations and market research; the new professions that had 'developed their skills and pretensions in order to serve men whose God is the Big Sell'. Mill's assessment of the designer's role would likely not have been a shock to the audience gathered at Aspen or to the readers of *Industrial Design*. Sociologist Vance Packard's book, *The Hidden Persuaders*, had recently been published (1957), popularizing a movement against the so-called 'manipulation' professions, including advertising.[10] In 1970, Packard's book *The Waste Makers* had focused its critique on the concept of planned obsolescence, putting the industrial designer at the centre of its attack, within wider debates about cultural value and the morality of consumerism.[11] In bringing these debates inside its tents, Aspen offered a space in which the hegemony of the corporation and big business could be effectively managed. By contrast, professional organizations, particularly the SID and SIA, were ominously silent.

In her detailed history of the IDCA, Alice Twemlow has argued that Aspen also provided a notable setting for transatlantic exchange between designers in Britain and the US.[12] This was most obviously expressed in the role played by Reyner Banham, who used it in part to provoke a more commercially oriented dialogue about design than was possible at home in Britain.[13] The divergent identities of the industrial designer in Britain and the US centred on a simplified and exaggerated distinction between the anti-commercial versus commercializing impulses of the two professions. This can be read in the speeches and reports for the 1960 Conference, 'The Corporation and the Designer; an Enquiry into the Opportunities and Limits of Action for Innovators in the Twentieth Century'. This event highlighted the major disparity between the two professional cultures and, crucially, positioned the corporation as the mediating agent in this distinction.

In something of an odd choice, British naval historian C. Northcote Parkinson delivered the keynote address. Parkinson had recently become a well-known figure in business management critique, through the publication of *Parkinson's Law: the Pursuit of Progress* (1958), a book that brought him great personal fame on both sides of the Atlantic.[14] 'Parkinson's Law' referred to the truism that 'work expands to fill the time allotted to it', and his theory was developed through experience of administrative bureaucracy in the context of the Second World War. It was immensely successful and widely interpreted to fit any number of professions, structures and working identities, from the American middle-class housewife to public welfare. The argument was interpreted in the British media as a criticism of American management techniques that had been apparently 'exported' to Britain since the post-war period. In this sense, his work spoke across the apparent 'psychological gulf' between British and US professional cultures.

While Parkinson might have been a controversial figure in business circles, the content of his speech confirmed his general suspicion of the image and identity of the artistic professions, as he presented a case in defence of the Victorian 'professional ideal'. Once again, the image of the bearded designer came under attack as Parkinson differentiated between those who wear the 'grey flannel suit' and those who 'prefer the beard, the beret and the sandals' (incidentally, offering a prophetic insight into the opposing, and later combustive elements within the Aspen audience itself).[15] Nevertheless, as a businessman, his speech offered limited knowledge or insight into the issues troubling the design profession. Perplexingly, his proposition to the design community assembled at the IDCA was to 'follow a creed!'

> My counsel to the designers of the world is to make art a discipline, train their successors in an accepted tradition, set their professional standards and establish their professional examinations. Each December, at least, one member of the professional body should be expelled for producing the worst design of the year. And what of the rebels, the eccentrics, the deviationists? By establishing an accepted tradition, you will do them the greatest possible service. You will have given them something against which to rebel.[16]

This advice, seemingly given without much awareness of the development of professional organizations in the field of industrial design or their attendant problems, echoed the language of professional career guidance in British design circles, including that of Dorothy Goslett and Cycill Tomrley. It seemed to point designers in precisely the wrong direction as they faced the declining significance of the professional ideal in modern society. Responding to the piece, which Parkinson later published in his own *Newsletter*, in private correspondence, Henry Dreyfuss and John Vassos were dismissive of its value, which did not seem to make much sense to them.[17]

In writing about Aspen, design historians have hitherto focused on the dramatic tensions that played out at the end of the 1960s, as the Conference descended into generational conflict,[18] but the programme booklet for the 'Corporation and the Designer' (1960) presents the event in a much more conservative light. The Conference was split into three cycles that year: 'the identity of corporations', 'the identity of design' and 'the future of design in a technological society'. Willy de Majo, President of the SIA, was an adviser to the Executive Committee Council. Speakers included businessmen, designers and sociologists from both sides of the Atlantic and on both sides of the argument. Leslie Julius, managing director of British furniture company Hille, spoke about his preference to work without a contract on the basis of a 'gentleman's agreement', which was the nature of his employment of Robin Day, a long-term consultant to the company. In his paper, US designer Eliot Noyes explained that the relationship between the designer and the corporation in the US, 'in a society where national welfare contributions rest on the corporation', was fundamentally different to that of Europe or Britain.[19] For Vernon Welsh, an American independent consultant, the 'future was bright for men who design corporations', as he presented something of a eulogy for the romanticized ideal of the Consultant Designer, describing 'men with inherent design sense, film sense, theater sense, pattern-sense, who seek elegance, imagination, daring and style'.[20] Sir Paul Reilly, Chief Executive of the CoID, made one of the lengthiest interventions and presented the context of the cold war as the ultimate challenge for the industrial designer, pledging his allegiance to the values of 'good taste'. Revealing the social basis of design reform discourse, he said 'Deciding classes have lost their cultural nerve.' The designer had to renew his responsibilities to 'his client, his market, to his times and to himself'.[21] If this was the design elite's response to the public critique of capitalism, industrialization and the new professions, it was a very timid one. While it would later provide a platform for visceral and insurmountable dissent, this particular Aspen conference revealed the reluctance – or perhaps inability – of the 'founding fathers' of the industrial design profession to face up to longstanding problems. It precipitated the escalating crisis within the profession as designers struggled to articulate social relevance and value and move beyond the tenets of good taste, professional expertise and planned obsolescence.

Later years of the Conference saw the effects of this generational ignorance play out in spectacular form. The formation of the Conference Board, which drew upon top levels of American management consultancy and industrial design, presented an obvious culprit at which disaffected students could take their aim. Twemlow's captivating analysis of the 1970 event emphasizes the importance of the generational divide in the radicalization of the Conference, where two contingencies of design profession – often manifest in the generational hierarchy of professor and student – were unable to talk to one another.[22] Where students had initially been invited to attend free of charge (in exchange for voluntary work at the event), things changed when 'they attended in greater numbers and were asked to pay'.[23] A remarkably contemptuous Aspen Board remarked in 1969 on how the students 'lacked direction: it is impossible to predict much about them or their attitudes'.[24] Reyner Banham later reflected, 'once a distinctive student culture began to emerge, taking neither professionalism and professional status seriously nor for granted and began to replace the deferential boy-scoutism of students at earlier Aspens', the event deteriorated into dissent. Banham reflects on the culture clash between the business attendees and students in particular, which fed a 'well-nourished paranoia about long hair and bare feet',[25] a paranoia that had been equally expressed in the pages of the *SIA Journal* some twenty years earlier and in the pages of *Vogue* in 1939, as discussed earlier in this book. Driven by the politically informed 'refusal to work', Italian designers from groups including Superstudio and their 'radical architecture comrades' drove the most publicized and memorable critiques of the design profession at Aspen.[26] In 1970, partially as a result of the active participation of this radical, 'new generation' of designers and design thinkers, the Conference came to function as a site of protest and controversy. Much of this debate addressed questions of value, work, ecology and the environment that were external to the profession and driving social change beyond the limits of design.

An image from the 1970 IDCA Student Handbook, an unofficial publication from the conference, with a 'sculpture of junked cars and appliances, painted white and assembled in Aspen by students from Northern Illinois University under their professor Don Strel, 1969'[27] (Figure 5.1), presents a striking comparison with the image presented in Chapter 1 of the polished, suited designers eagerly working on the Ford exhibit at the New York World's Fair thirty years previously (see Figure 1.5). Adhering to the same basic sculptural form, they show industrial design at opposite ends of the professionalization process, from its sincere intentions as a 'serious new profession', to radical disillusionment and abandonment of the professional ideal.

Professionalism and the designer

Design was not the only profession to undergo existential crisis in the 1960s. Architects in Britain also underwent a period of reflection on their identity as professionals, well documented in *The Architect and His Office* (1962), a report published by the Royal Institute of British Architects (RIBA). The impetus for this report came from a crisis of professionalism within the RIBA, in 'an increasingly competitive and anti-Professional world', as summarized by the architect Owen Luder in *The Architect and Building News*:

> One basic fact as a profession we seem to be unable to accept is that many of our inbuilt defences against the outside rules, excellent rules for regulating our own affairs, such as the Code of Conduct and fixed fees, are fast becoming a liability rather than an asset.[28]

As Luder made clear, one of the most obvious reasons for this was the emergence of a more competitive and dynamic economic market:

> Architects brought up in an atmosphere of pure professionalism are perhaps inclined to fool themselves into thinking that clients and the public put as high a value on professional independence and unlimited liability as they do. The fact that package dealers who offer neither of these two qualities are growing in size and influence every day indicates the opposite may be the case.[29]

Crisis of professionalism 159

5.1 Cover of Student Handbook, International Design Conference in Aspen (IDCA) (1970). Courtesy of the Paul J. Getty Research Institute. Copyright Don Strel, with kind permission of Don Strel Estate.

Professional organizations were being forced to respond to wider critiques and cultural change that extended beyond the limits of their own field of practice. As it moved into the next decade, one of the most difficult things the SIAD had to face was the devalued and declining significance of professionalism as a sustaining working ideal. In 1965, SIAD President

John Reid admitted that 'to talk of being professional today is to invite the accusation of Victorianism or fuddy-duddyism .... Professionalism is not a Code of Ethics. It is a way of life. A code of ethics is not just a set of antiquated rules which say "thou shalt not" to anything and everything that you want to do.'[30]

Speaking at the SIAD Annual Conference in 1968, President Milner Gray told the audience, 'We live in an age in which the winds of change seem to have reached gale force, so fast do they blow.'[31] Gray had been urged to take on the Presidency of the Society again that year, to steer them through this turbulence. Responding to this overwhelming context of change, in 1968 the SIAD commissioned consultant James Pilditch to conduct a report entitled *Professionalism and the Designer*. The aim was to evaluate the concept of professionalism, and the relationship of the designer with other professional bodies in government and industry in Britain and internationally.[32] At the beginning of the report, the authors Michael Middleton, Peter Lord and James Pilditch set out the 'new order of existence' in which contemporary designers worked. Positioning the designer as a 'person who offers society alternative futures', they pointed to a very different set of imperatives for the design profession, opening up a new context in which to consider the designer's role. They remarked that 'the gap between professional morality and "commercial morality" has narrowed markedly, so that in many directions, for much of the time, public companies now behave no less scrupulously than would a professional practice.'[33]

The decision to commission Pilditch to write the report was in itself indicative of the scale of change within the British design profession as it moved from a position of pure professionalism to a more commercially flexible attitude. As discussed in Chapter 2, Pilditch had gained a reputation for being commercially oriented in his view of design work, a reputation he proudly documents in his books, *The Silent Salesman* (1973), *Talk About Design* (1976) and *I'll be Over in the Morning* (1990). Nevertheless, he was also essentially sympathetic to the aims of the SIAD. In *The Silent Salesman*, a section entitled 'How to be a good client' essentially repeated the rhetoric of the SIAD in its publication 'Working with your designer', instructing the client on 'good professional practice', including the bar on 'free-pitching' and the importance of contractual agreements in fee negotiations and work completion.[34] While the *Professionalism and the Designer* report was a critical moment of internal self-reflection, the Society chose to take opinion only from 'essentially sympathetic bodies and contacts', including Paul Reilly, Director of the Council (1959–1977) and Honorary Fellow of the Society. With the exception of Pilditch, the report therefore reflected the outlook of the usual design circles.

Research for the report was conducted at central headquarters in London. Interviews were set up with past Presidents and a meeting arranged with the sociologist Geoffrey Millerson, who is quoted frequently in the report. In addition, the views of designers Barbara Jones, Alec Heath, Ernest Hoch, Jack Procter, Frank Height, Ronald Dickens and Andrew Renton were sought. These Fellows of the Society covered a wide range of disciplines. A working questionnaire was circulated with the intention of gauging their general view on questions including: 'Is there a definition of professionalism which is generally accepted? If not, why not?' and 'If the two main aspects of professionalism are held to lie in a) competence b) behaviour, are the two standards generally adhered to? If not, is it possible and to what extent is it desirable to "police" either?' Only one response to the questionnaire survives in the Society's archive; that of Lily Goddard, textiles designer, who simply stated, 'To any artist, professional codes can only be a guide, and changing worlds a background to his ideas', a poetic statement that revealed the essentially artistic identity of many of its formative members, who continued to practise exclusively as illustrators or graphic designers.[35] Reilly's feedback to the Council, obtained at an informal meeting with Michael Middleton at the Arts Club in London was, unsurprisingly, in line with the Society's own view: that the key to professional behaviour is 'competence, honesty, integrity'. Taking notes on the meeting, Middleton stated:

> Sir Paul made the important point that attitudes to advertising, like morals, tended to move in cycles. What was abhorrent 10 years ago is today acceptable. He foresaw that in five years or so, we would find ourselves in a more puritan society with more puritan standards than might exist today.[36]

Pilditch's notes confirm that advertising and self-promotion continued to be major issues for designers. Reilly acknowledged the value of PR and promotional activity as a necessary tool in the Consultant Designer's professional equipment, a radical departure from the staunchly anti-commercial approach of British design promoters in the post-war years. He said, 'on promotion too, Sir Paul felt that there is everything to be said for first class designers to have good PR officers on their staff'. In answer to the question of why PR was permissible over advertising, he said that with PR work, there was an 'element of professional judgement involved', a statement that echoed words expressed by the SID in the US some two decades earlier.[37]

The questionnaire circulated to membership was focused on key issues that arose from these preliminary discussions at top level. There were 580 respondents to the questionnaire, one-sixth of the total membership (3,035) at the time. The results of the report were illuminative on the understanding and value of professionalism to practising designers in 1971. In general, members were critical of the relevance of the existing Code of Conduct. Although 80 per cent of the Society preferred to describe design as a profession than a commercial service, 93 per cent prioritized 'competence' above all other 'ingredients' of professionalism. These were identified as: 'honesty, competence, confidential relationship with client, reliability, responsibility to community, sense of service, objectivity, responsibility to fellow practitioners'.[38] Therefore, overall, the behavioural qualities that had underpinned the Society's Code of Professional Conduct since 1945 were now deemed less significant than technical ability. The need for a Code of Conduct was accepted by 93 per cent of the membership, though fewer were 'convinced that a unified Code is fully possible over the whole spectrum of design, or believe a Code to be fully applicable to the salaried practitioner'. In drawing conclusions from the questionnaire's findings, Pilditch urged the SIAD membership to reconsider its identity in relation to the 'older professions' it had so closely tried to mimic. He wrote, 'there are justified differences between the professions … the jobs of lawyers, doctors and architects are lumped together with the newer professions, but their jobs and responsibilities differ totally'. The report also noted that in the nineteenth century, sociologist Alexander Carr-Saunders had noted that it was 'impossible to accept the claims of advertising to professional status' on the basis of its role in 'producing half-truths and bogus scientific terminology to sell products'. Today, however, 'this argument might be outdated, as the terms and limits of the professional ideal, enshrined in the Code of Conduct, required revision across almost all professions to fit the changing conditions of today'.[39]

Pilditch and Stuart Rose, CSD President, redrafted the Code of Conduct on this basis of the report's recommendations, but their revisions were rejected by the SIAD Council on the grounds that ' a member may issue to the press illustrations, factual descriptions of his work and biographical material for publication', bringing tensions about advertising and self-promotion to the surface once more.[40] In a further redraft, the tone and format for the Code was completely revised and renamed 'Declaration of Professional Behaviour', a softer title that hinted towards the limitations of applying a Code of Conduct in the design profession. Rose and Pilditch agreed in their report that the composition of the SIAD was 'totally different' compared to that for which the Code was originally formulated. The Commission offered the recently accepted definition of industrial design at the ICSID, formulated by Tomás Maldonado.[41] It was a rare instance of the Society looking outwards, attempting to incorporate internationalism to its national organizational view.

With a rather ironic sense of timing, the second edition of Dorothy Goslett's book, *The Professional Practice of Design*, was published in 1971. Although *Professionalism and the Designer* and *The Professional Practice of Design* might sound like complementary titles published in the same year, in actual fact they presented two opposing attitudes to professionalism in design. Goslett's book reaffirmed its close interpretation of the SIAD's concept of professionalism, making it outdated almost as soon as it was published. The partisan nature of its advice was noted in a review in *Design* magazine:

> The SIAD is referred to throughout as if there were no possible doubt as to the desirability of all designers becoming members. There are a number of members (and I am one of them), who find among the SIAD's recommendations a lot with which to disagree, sufficient in fact to deter them from joining (or, as in my case, to persuade them into resigning). It might have been better, while still quoting from the SIAD, to have been less partisan.[42]

It was clear that a firmer voice of dissent against the SIAD was building. Goslett's statement that she had written the book under the instruction to 'tell young designers how to behave and what to do', as described in Chapter 3, was appearing increasingly out of touch.

Pilditch was direct in his observations about the limited hold of this gentlemanly professionalism in contemporary design practice, stating, 'there is nothing unprofessional in aiming to make a design business profitable'.[43] The Declaration of Professional Behaviour was now structured into three sections to address three main audiences, 'Responsibilities to Society at large'; 'The designer's relationship with fellow practitioners' and 'The designer's relationship with fellow professionals'.[44] The SIAD Council again rejected the newly drafted document in 1974, reporting a 'real concern that the present document doesn't allow either for adequate control of the professional behaviour of its members'.[45] It was defeated at the annual general meeting of the SIAD in 1974, against a rising defence of the standards laid down in the Code of Professional Conduct. In the context of this controversy and crisis, the Society again requested individual opinions from a number of Fellows and associated and prominent members in 1973, again from those the Society considered broadly in sympathy with its aims. This included Dorothy Goslett and General Consultant Designers Willy de Majo, Jack Howe, George Fejer and Peter Ray. Almost unanimously, they were against the new draft and, unsurprisingly, one of the strongest reactions came from Dorothy Goslett:

> Totally misjudged document. Puts Society at a stroke outside the pale of all other Professional Societies in conceding that members may advertise, tout for clients and provide free services in the hope of receiving subsequent work. Panders to members whose aim is highly profitable commercialism without professional integrity. Couched in vaguest, most apologetic and woolly-minded phraseology … Will take members out of status of truly professional people and put them in a category of commercial artists. Big Boys will be able to afford full pages in the glossies while young struggling, free-lance only manage a couple of lines in the local weekly.[46]

Drawing attention to the distinction between a 'professional person' and a 'commercial artist', Goslett makes clear the social motivations behind professionalization in design. Similar concern about how design would be seen as a profession was expressed by architect J. B. Fuller, who wrote to the Society in concern that the SIAD would be seen as a 'poor relation of the RIBA'.[47] Misha Black, who had been actively involved in the exportation of the SIAD model of professional conduct in the ICSID Constitution, returned a spoiled ballot for the 1974 Declaration. He explained to the committee, 'I agree that we must all promote our profession, but I cannot believe that the specific permission to undertake paid advertising is either necessary or desirable ... I cannot support a code which equates, in this one respect, professional conduct with commercial salesmanship.'[48] Fellow Consultant Designer, Willy de Majo, was similarly frustrated, stating that he feared it would make design a 'poor relation', again referring to the image of the profession in the eyes of the architect, a continual refrain in British design. De Majo criticized the use of the word 'Declaration' in particular, which, he argued, gave the sense of 'trying to apologise for wanting to be professional, while not having the courage of our convictions'.[49] Jack Howe wrote, 'Document gets worse every time it is revised. Latest edition is vague, undecided and contradictory. Based on pious hope rather than professional authority and would make the Society a laughing stock.'[50] For these men, the Society was moving too far from its traditional base of 'pure professionalism'.

Recognising the need to obtain opinion from outside the SIAD, Chief Executive Geoffrey Bensusan addressed letters to Terence Conran ('a designer who advertises successfully') and textile designer John Tandy, inviting them to take part in a debate at the Society over the issue of advertising and publicity.[51] Neither attended. Tandy replied to decline, stating, 'My feelings on the subject are well-known.'[52] A new proposal, submitted to the Council by Bensusan, indicated the desperate lengths to which the Society would now go to maintain some grip on the identity of the professional designer. Bensusan proposed an extremely thorough and detailed list of every imaginable method of self-promotion – including Christmas cards – and split these into categories of 'freely permitted', 'with safeguards' and 'prohibited'.[53] It was not accepted by the Council. Instead, the Code of Professional Conduct, which was revised for the second and last time in 1975, did permit advertising, under a long list of defined clauses. It was increasingly difficult to find a middle ground between the views of professionalism held by those outside the SIAD and the older generation of members inside it. The document was full of complicated compromise. Clause 9.7 read, 'under no circumstances shall members seek to disguise paid advertising as unsolicited editorial content, nor by employing the services of another to promote their own services, shall they seek to hide the true source of that promotion. Members are advised to refer to the British Code of Advertising Practice.' The redraft also conceded that the Society's determination to be exclusive had inhibited its interaction with the cultures of professionalism that had developed beyond the limits of the SIAD, including other institutions. On the subject of design competitions, it stated:

> There are occasions when, in the interest of informed design judgement, it is better for the Society to be represented by a number on a judging panel than to have no say in the judgement, even though the Society may not have been able to give the competition its official blessing.[54]

As Gordon Russell stated, 'the idealistic motives of the British design tradition were being called into question' as the design profession moved into the next decade.[55]

The IDSA: a new code of professional conduct

The merger between the IDI and SID, brokered by Henry Dreyfuss and Raymond Spilman, responded to the building sense of crisis within the US industrial design profession and an awareness of the declining significance of professional organizations internationally. Ray Spilman wrote to William Renwick in 1963 to say that he was concerned 'that both Societies will die if we do not get together'.[56] Rumours of the proposed merger, which had been circulating within the profession for some time, were not welcomed by everyone. For Walter Dorwin Teague, founder of the SID, it signalled only a weakening of his original aims, and a devaluation of the industrial designer's exclusivity within professional design discourse:

> It is deeply disturbing to me to have IDI constantly coupled with SID as if they were Tweedledum and Tweedledee. Since the beginning, some of us have struggled to maintain SID in a unique position, not from any snobbish inclination but because we hoped to make it an exclusively professional Society of industrial designers in which membership is an honour ... It seems to me our position has been crumbling around the edges for some time ... I hope the superior position of SID can be salvaged ... otherwise, why bother? I don't like societies or clubs as such anyhow.[57]

As he clearly articulates here, for Teague, the merger of the two societies represented a dilution of the culture of elitism around the identity of the industrial design consultant he had worked hard to protect. The rather surprising and revealing statement, 'I don't like societies or clubs as such anyhow', is further evidence of the highly pragmatic view Teague had always taken of professionalism and professional organizations.

Interventions were needed to calm discord between the societies in preparation for the merger and the process had to be delicately managed by Dreyfuss, Spilman, Vassos and others. In 1963, writing to Vassos, Jon Hauser said, 'we are at the threshold of a new era for professional designers and any break in our ranks could seriously impair the progress so important at this time'.[58] Edgar Kaufmann, Jr, who had chaired the discussion on professional identity in design at MoMA in 1946, described in Chapter 1, wrote to congratulate the two organizations on their merger and for getting 'off to a good start under the best auspices'.[59] The incomplete nature of professionalization was noted by the Society's early declaration to 'make industrial design a vital profession', an aim that had been stated by the SID in 1944. Anticipating the merger, Spilman was appointed to redraft the ASID Code of Ethics, something he referred to as the necessary 'raising of professional standards'.[60] During this process, Spilman consulted a great deal with founder Henry Dreyfuss. Dreyfuss wrote to Spilman in 1963, advising him to exercise caution in the raising of these standards, drawing attention to a significant contingent within the SID membership who primarily identified industrial design as a business and not a profession. This group included, he stated, 'Loewy, Van Dyke Assoc., Cushing and Nevell and smaller offices that included John Cuccio, Roger Singer and others'. Dreyfuss urged Spilman to find a way of revising the Code that would integrate rather than isolate this group:

> As the years go on, we cannot afford to retire to an ivory tower. Just think how industrial design has changed over the past ten to fifteen years. Because a man makes a success out of a profession, I don't think you can just put him down as a businessman.[61]

As Dreyfuss indicates here, the IDCA had helped to reshape, redefine and integrate the relationship between business and professionalism in US industrial design. His letter also identifies the essentially individualist nature of US industrial designers and the essential failure of any professional organizations to unite them, stating that these men would 'just join another group'. Rather, he said, an effort should be made 'to try to influence these people so they will accept the mantle of a profession, live by the Code of Ethics and remain part of our Group'.[62] As Dreyfuss correctly intimates here, professional identity in industrial design involved a complex negotiation between lofty ideals and economic incentive. The material value of being seen to be professional, as it was articulated to the SID in the post-war period to industrial designers working in New York, no longer stood. As Dreyfuss put it to Spilman, tightening the ethical stance of the Society no longer made good business sense: 'I want to be sure we just don't retire to a nunnery.'[63] Like Pilditch and his contemporaries in Britain, Dreyfuss was starting to feel the limits of professionalism as a working model for practising designers in an increasingly competitive business world.

The IDSA's seventh annual symposium was aptly entitled 'The Professional Challenges of the 60s'.[64] Membership was dwindling. A list of new applicants in July 1965 was described by Henry Dreyfuss as 'not a happy showing', with only six in New York City, two in Detroit, two in Los Angeles and nine in Chicago.[65] An undated memo circulated within the IDSA Board of Directors said that the IDSA should appeal to groups 'not proportionally represented, specialist, corporate designer, educator, consultant staffs and in general the younger designer'.[66] By 1965, the IDSA membership survey recorded an even split between consultant and corporate designers (180/182), but the respondents to the survey decidedly felt that the future of the profession lay in working as part of a corporation, rather than as a consultant.[67] Feedback from the survey was generally rather negative, with young members reporting on a 'stale attitude', urging for activities in the IDSA that involve 'anything but contemplating our own navels year after year'.[68]

The IDSA's daily business was increasingly taken up with ego-driven disputes between industrial design consultants. In June 1964, the SID suffered the public embarrassment of the resignation of perhaps its most well-known designer, Raymond Loewy. He tendered his resignation in February 1964, citing 'gradual deterioration of ethical relationships among members which I consider to have reached danger point'.[69] Loewy was referring here to a dispute with consultant Brooks Stevens. In an article in the *New York Times* on car stylists, Stevens had been given full credit for all Studebaker products except the Avanti, but Raymond Loewy was upset because the piece did not state that the original car was of his design. Stevens refused to apologise to Loewy, claiming that it was not within his ability to control what was published in the article. In an angry reply to the SID grievance committee, Stevens stated that 'the matter with Loewy has become exaggerated beyond my degree of patience'.[70] Meanwhile, Loewy was offended that the SID did not immediately take action against Stevens, instead being asked to substantiate comments from his letter. In a letter to Renwick, Don Dailey of the SID pointed out Loewy had been guilty of unprofessional activities (advertising through a sales brochure) earlier in the year, adding that he had a sense that both men 'were concerned with their own image rather than the welfare of the profession'.[71]

Climate of discontent

In 1967, the IDSA led an educational seminar at the Plaza Hotel in New York, 'to objectively examine industrial design education from the standpoint of improvement leading towards a better design profession'. Dave Chapman, reporting at the event, described a 'big change' in education, which he related to changes in youth culture and identity more broadly. Students 'shy away from big offices', he said. They wanted to '"fight the establishment", only 20% are married, and they are pessimistic about the future … Students want purpose now, not only an income.'[72] Dreyfuss wondered if the academic world was 'scorning' the design profession.[73] Arthur Pulos, head of industrial design at Syracuse University, said, 'We need a counter-revolution' to face the 'anti-design' attacks. An undated memo in 1968 stated that the IDSA should appeal to groups not proportionally represented, 'specialists, corporate designers, educators, consultancy staff and in general the younger designer'.[74]

In May 1969, a report was submitted to the IDSA Board of Directors entitled 'Climate of Discontent', summarizing in a provocative list of nine the perceived 'problems' of the IDSA. Among them, it was stated that the IDSA 'has an exclusive, fraternity image', not representative of the profession; is 'consultant oriented'; has a 'poor relationship with both the public and student populations'; has 'inaccessible entrance requirements'; and is 'overly concerned about the quality of our membership'. Summarizing, the writer stated, 'we don't have a good enough answer when someone says, "'what's in it for me?"; "what are our specific accomplishments?"'[75] Arthur J. Pulos attended a meeting of the Los Angeles and San Francisco chapters in May 1969 and reflected on 'much frustration and discontent, especially among younger members'. The same year, the Society conducted a membership survey, the feedback from which pointed in much the same direction. In a revealing statement, a significant majority felt membership of the Society had no influence on their ability to attract clients. The greatest fault, some said, was the inability to achieve licensed professional status alongside the architect and the AIA. A significant number felt that the industrial designer had a lower reputation than other professions. In May 1969, a memo from F. E. Smith to the IDSA Board of Directors, entitled ' Climate of Discontent', registered a professional crisis within the Society. The memo outlined a long list of complaints that had been heard across the country, particularly among younger members and at student meetings in Los Angeles and San Francisco. These included an 'exclusive fraternity image', and the accusation that the IDSA was 'consultant oriented. Of our 29 board members, only 6 are corporate staff employees; only superficially recognize design students'. More existentially, the memo ended with 'what are our specific accomplishments?'[76] Reflecting on this memo and a recent meeting he had attended of student members in Los Angeles, Arthur Pulos conveyed his feeling that the Society should turn to face these challenging criticisms, namely addressing the issue of 'preventing design obsolescence', which should now be a major goal within the Society.[77]

Creativity

As the limits of professionalism and its value in design came into question, creativity became a more attractive alternative creed for industrial designers in Britain and the US. Speaking in March 1959, Raymond Spilman, who drove the evaluation of professionalism in the ASID before the merger, gave his thoughts on professionalism to the New York chapter of the AIA, stating that the 'crux of the emotional and professional problem is who is to control the execution of creative effort, the businessman or the professional':

> At this moment, there is a growing group of creative businessmen in this country – like Walter Paepcke of Container Corporation of America, William M. Stuart of Martin Senour Company, John D. Rockefeller III, John Hay Whitney and many others – leading and guiding American industry into new channels of thought and expression. These men have challenged the creative profession to contribute something new and different and in their own image, rather than in the image of the businessman. In effect, they have said, 'show ourselves, not as we believe ourselves to be, but as you think we should be'.[78]

As Spilman articulates here, the move towards integration, the origins of which were described in Chapter 2,[79] represented more than a structural reorganization of the designer within the corporation. It was underpinned by a central idea of creativity – a guiding principle that was reshaping the way US businesses positioned themselves in the cultural economy. Spilman argued that integration had facilitated a new type of professional: the 'creative professional'.[80]

Here, Spilman confirms the central role of the corporation in shaping professional identity in industrial design in the US. The venture of corporations into aesthetics and design had ushered in a new era of professionalism, and Spilman argued that the relationship between professionalism and business had been altered through the exchange. According to Spilman, this was a place where the US was leading the way.[81] This turn towards creativity as a more flexible working culture was also embraced by graphic designers and art directors in Britain, who had come to see the British values of professionalism in design as stale and outdated. Writing in 1961, James Pilditch dedicated an entire chapter of *The Silent Salesman* to creativity, defined rather obtusely as 'not just an answer to the problem, but a solution plus something else – an indefinable spark that will life [sic] the whole project from the rut of every day'.[82] Pilditch quoted from Raymond Loewy Associates, 'It is only creative ability which can take research, facts, statistics and convert them into an object, a shape, a building, which can command cash from consumers: it is in the end creative ability which creates a sale', laying bare the value of commercialism in the function of design practice.[83]

Contrary to the predictions of C. Northcote Parkinson, who advised a tightening-up of professional organizations in design described at the beginning of this chapter, the SIAD's rigid approach to professional behaviour shaped a rebellious attitude among younger designers, who had been lectured to by the older generation. As graphic designer Ken Garland put it:

> On the whole, those people who were my friends and colleagues ... they included Alan Fletcher, Colin Forbes, Derek Birdsall ... A whole slew of us who were great enthusiasts for the business of graphic design, the activity of graphic design, were a bit sniffy about the profession. We thought it looked a little jumped up, like it was trying to ape the professional codes of architecture or engineering ... I also thought there was a certain freedom about not being a professional and I think my colleagues probably thought the same way.[84]

Garland's statement here about the 'freedom' of not being a professional articulates the cultural distance that had now moved between the notions of professionalism and creativity. Garland co-founded the Design and Art Directors Club (D&AD) in London in 1962, importing the model of the US Art Directors Club. Its main activities concerned competitions and awards, with the production of an Annual each year. The generational divide between the D&AD and SIAD was referenced in an article in *Creative Review* magazine in 2012, in which it referred to the SIAD as the 'bowler hat and umbrella brigade', language that captures the enduring codes of masculinity that continued to drive identity in design at every turn.

First things first

In 1962, the SIA, led by FHK Henrion and Stuart Rose, held a meeting at the Institute of Contemporary Arts, London, entitled 'Why you should join the SIA'. This meeting aimed to bring together a discussion between existing members and Fellows of the SIA and the increasing numbers of designers who had chosen not to join. Garland remembered the meeting vividly in an interview in 2013:

> I almost decided to leave, I was at the back, ready to go, and normally I sit at the front. I stayed on until the end and I rather lost interest in the prophetizing of the older generation and I started to write what I really thought about design, and I wrote what later became called the First Things First Manifesto and at the time, I didn't think much of it ... a man called Stuart Rose, a nice guy incidentally, said, 'has anybody got anything else to say?' and I thought, 'well why not', so I raised my hand and I read out this thing, which as I read it became more declamatory and it turned into a manifesto during the reading ... It added some life to the meeting and afterwards people asked me, 'are you going to publish this? I invited some people to sign it, some said yes, some said no, I was quite impressed by the ones who said no, and my friend Herbert Spencer, who was half way between the older and younger generation said no, for example.[85]

The 'First Things First Manifesto', as it subsequently became known, was published and circulated in 1963, and continues to be a touchpoint for ongoing debates about professional and ethical values in graphic design.[86]

Of the many fascinating things to take from this moment of ethical reckoning in British graphic design, perhaps the most striking is the thirst for ethical standards as alternatives to the existing professional codes offered by the SIA and one which, notably, dealt directly with the relationship between the designer and the public. In the years that followed, there would be many critical voices expressing the need to abandon the professional ideals of the 'first generation' of industrial designers. Garland was, in actual fact, calling for something simpler and purer – an alternative value system. When he published *First Things First: A Manifesto*, a widely

Crisis of professionalism

5.2 Ken Garland, *First Things First: A Manifesto* (1963). Copyright Ken Garland, with kind permission of Ken Garland Estate.

circulated document with more promotional value than the SIA Code of Conduct, it is striking to note how he arranged the principles in a codified format that refashioned the SIA Code. The popularity of this code made two things about the SIA clear: first, that the graphic design constituent of the Society continued to put its views on professionalism under most scrutiny; secondly, that there was an appetite for rules, ethics and codes that moved outside and beyond the ideals of pure professionalism.

By the end of the decade, the professional ideal had begun to reach its limits as designers challenged and moved outside professional norms and behaviours that had been laid down by professional organizations. In 1964, another SIAD member, Anthony Perks, wrote to the *SIA Journal*:

> The Society is drawing into a polite and senile old age. Some day, we shall have premises of our own where aged designers can sit in old leather Hille armchairs and bore other aged designers with stories of the waste-bins they designed for the Britain Can Make It exhibition. When Saul Bass visits this country, it is to another Society that he lectures, not the SIA, but the STD who run a series of mostly stimulating lectures through the winter. Those of us who have no dinner jackets get little from SIA nowadays except an affix at nearly two pounds per letter per annum.[87]

Perk's reference to Saul Bass, an American designer of international fame, reflects the endurance of the US as a cultural signifier through which British designers measured their own professional success. In 1968, illustrator and president of the SIAD (1951–1952) Lynton Lamb was invited by Peter Lord, SIAD President, to express his thoughts on the SIAD Commission on Professionalism. Lamb was surprisingly forthright, suggesting to Lord that the Society's problems with professionalism had taken root in the formative years, when there was also a 'revolt against a sort of Curia of elder designers, who gave the appearance of wanting to use the Society as a backing for their own professional ambitions, but who had found that for this purpose it needing dusting down and dressing up in a more respectable disguise'.[88] He explained, prophetically:

> I have always thought that we chose the wrong profession to copy, and that the RIBA would take us particularly up the wrong creek. We have been behaving like this, that and the other manqué in a national life that has infallible instincts for spotting the outsider. It seems that we spend too many of our resources trying to buy prestige instead of earning it.[89]

They were searing words for a society seeking to reinvent its professional self-image.

Conclusion

All professions go through moments of crisis. However, as this chapter has explored, many of the struggles professional design organizations faced related to problems first encountered in their formative years that had been percolating under the surface ever since. This includes the struggle between the identity of the artist versus the professional; the role of advertising and self-promotion; tensions between consultant versus corporate designers and an unresolved status as a new versus old profession. Whereas the first two chapters of this book illuminated essential differences in structure, identity and organization of the industrial design profession in Britain and the US, it is striking to note the extent of overlap within this chapter, as professions on both sides of the Atlantic faced almost identical questions of professional crisis and identity. By the 1960s, the discourse of crisis was driving the professions on both sides of the Atlantic, functioning as a system of meaning-making in a profession that lacked professional structure and status. While well-established tensions between commercialism and anti-commercialism were at the surface, at its root, the crisis was more existential and more fundamental. It attacked the industrial designer from both sides – the negative moral effects of consumerism, along with attacks on the dull anonymity of the professional ideal. It would be difficult to argue that any organization in Britain or the US emerged stronger from the professional crisis of the 1960s. This is to some extent attributable to the shallow nature of their engagement with key issues, including social responsibility. For all of its soul-searching, questions of representation were notably absent from the dissenting discourse. Among Pilditch's recommendations in 1968, he added 'a better balance between men and women should be struck', a statement that considerably understated the profession's gender problems.[90]

Perhaps the thing that most troubled those within the SID and SIAD was the ease with which Terence Conran and Raymond Loewy chose to step outside the professional society and continued to function, to great material success and public visibility. It was clear to see that those designers who had never joined a professional society were unaffected by this decision. Similarly, William Goldsmith wrote to John Vassos on 27 April 1966 suggesting they should reconsider the election of Charles Eames and Eliot Noyes as Honorary Fellows, since 'their contribution to design has been substantial but participation in Society affairs has been less prominent than others listed', a statement that conceded the limited impact of professional organizations on commercial success.[91] More broadly, a number of factors beyond the control of any national design organizations – cold-war politics; student unrest; the Vietnam War, the environmental movement and generational social change – placed the profession in a more complex political arena. The profession proceeded with a patchy and ill-defined relationship to the 'public', the 'user' and the 'social' throughout the 1970s. Between 1970 and 1980, the industrial designer underwent perhaps the most dramatic reinvention yet, as the call to social responsibility was pressed upon a profession previously focused entirely on its relationship to clients, corporate agendas and consumers. Industrial design was ill-equipped to respond to this new imperative, as designers and the professional organizations that claimed to represent them had never adequately defined the nature of their relationship to the public. The next and final chapter, which concerns social responsibility and the industrial designer, addresses these shifts directly.

Notes

1 Tomás Maldonado, ICSID Venice (1961), quoted in Tania Messell, 'Constructing a United Nations of Design: ICSID and the Professionalization of Design on the World Stage, 1957–1980' (PhD thesis, University of Brighton, UK), p. 71.
2 George Nelson, 'A New Profession?', in *Problems of Design* (New York: Whitney Publications, 1960), p. 51.
3 See David Edgerton, *The Shock of the Old: Technology and Global History Since 1900* (Oxford: Oxford University Press, 2007).
4 Vassos discusses the early years of Aspen in his interview with Ray Spilman: 'I guess I was interested in things being broader and international, and not being what I saw then as so blatantly self-serving. But perhaps they weren't. The dream has turned sour. Aspen is still very popular, they fill it every summer. It's a hell of a tourist attraction, but somehow it's fallen into a pattern where it doesn't mean

much any more. Most of us don't go.' Vassos also intimates the importance of 'Mrs Paepcke' in the early years of the conference, although this is not further developed. Interview, c.1970s, Spilman papers, taped interviews, SCRC.
5. Robert Gordon-Fogelson, 'Total Integration: Design, Business and Society in the United States, 1935–1985' (PhD thesis, University of Southern California, 2021), p. 93.
6. C. Wright Mills, The Man in the Middle', *Industrial Design* (November 1958), 70–5, press clipping in IDSA Archive, Obsolescence folder, SCRC.
7. Ibid.
8. Twemlow, *Sifting the Trash*, p. 13.
9. Ibid.
10. Vance Packard, *The Hidden Persuaders* (New York: David McKay, 1957).
11. Vance Packard, *The Waste Makers* (New York: David McKay, 1960).
12. Twemlow, *Sifting the Trash*.
13. Banham was critical of the 'Good Design' principles of the CoID in particular. See Eleanor Herring, *Street Furniture Design: Contesting Modernism in Post-War Design* (London: Bloomsbury, 2016), pp. 164–9. See also Reyner Banham, *The Aspen Papers: Twenty Years of Design Theory* (London: Pall Mall Press, 1974).
14. The original theory was first printed in *The Economist* (UK) and *Harper's Magazine* (US).
15. Alice Twemlow describes the two factions of the Aspen conference in great detail in *Sifting the Trash*, pp. 104–7.
16. 'The Corporation and the Designer', 10th International Design Conference in Aspen (19–25 June 1960), Conference program, PGA.
17. Henry Dreyfuss to John Vassos (11 October 1965), Vassos papers, SCRC.
18. See Alice Twemlow, 'I can't talk to you if you say that: An Ideological Collision at the International Design Conference at Aspen, 1970', *Design and Culture*, 1:1 (2009), 23–49.
19. Eliot Noyes, Cycle II, 'The identity of design: experiences in corporations', Conference Program (1960), IDCA Archive, PGA.
20. Vernon Welsh, Cycle III: 'Possible Courses of Action leading toward more productive patterns for designers and Corporations', Conference Program (1960), PGA.
21. Sir Paul Reilly, IDCA Conference Program (1960), PGA.
22. Twemlow, 'I can't talk to you'.
23. Twemlow, *Sifting the Trash*, p. 104.
24. Ibid.
25. Reyner Banham, quoted in Twemlow, *Sifting the Trash*, p. 166.
26. Ross Elfline, 'Superstudio and the "Refusal to Work"', *Design and Culture*, 8:1: Work (2016), 55–77.
27. Twemlow, *Sifting the Trash*, p. 105.
28. 'The Architect and His Office, A survey of organization, staffing, quality of service and productivity presented to the council of the RIBA' (6 February 1962), National Art Library, V&A.
29. Owen Luder, 'The Writing's on the Wall', *Architect and Building News* (20 November 1968), Box 100, CSDA.
30. John Reid, Presidential Address 1965, 152/3 (October/November 1965), Box 200, CSDA.
31. Milner Gray, 'The Price and the Value', Presidential Address (1968), Box 200, CSDA.
32. Professionalism and the Designer Report (1971), David Pearson, personal archive.
33. Michael Middleton, James Pilditch and Peter Lord, 'Introduction' to 'Professionalism and the Designer' report (1971), David Pearson, personal archive.
34. James Pilditch, *The Silent Salesman* (London: Business Books, 1973 [1961]), p. 126.
35. Lilly Goddard, reply to Working Questionnaire, n.d., Box 100, CSDA.

36 Notes on informal meeting with Sir Paul Reilly, Friday, The Arts Club, n.d., Box 100, CSDA.
37 Ibid.
38 'Professionalism and the Designer', A working questionnaire, Box 98, CSDA.
39 SIA Commission on Professionalism (1970), p. 12, CSDA.
40 Draft 1972, Code of Conduct for the Professional Designer, Box 100, CSDA.
41 'Industrial Design is a creative activity, the aim of which is to determine the formal qualities of objects produced by industry. These formal qualities are not only the external features but principally those structural and functional relationships which convert a system to a coherent unity both from the point of view of the producer and the user. Industrial design extends to embrace all the aspects of human environment which are conditioned by industrial production.' Commission on Professionalism (1970), p. 16, CSDA.
42 Ian Bradbury, 'Review: The Professional Practice of Design', *Design*, 270 (June 1971).
43 Commission on Professionalism (1970), p. 16, CSDA.
44 Draft of Declaration of Professional Behaviour (1974), Box 101, CSDA.
45 Notes taken from special meeting to discuss the Declaration of Professional Behaviour, annotated by Pilditch (1974), Box 101, CSDA.
46 Dorothy Goslett, quoted in Summary of Responses to Declaration of Professional Behavior (30 October 1973), Box 100 CSDA.
47 J. B Fuller to the SIAD (25 November 1970), Box 101, CSDA.
48 Letter from the SIAD (25 November 1970), Box 101, CSDA.
49 Willy de Majo, Summary of Responses (30 October 1973), Box 100, CSDA.
50 Jack Howe, Summary of responses from members re: Declaration of Professional Behaviour (20 October 1973), Box 100, CSDA.
51 Geoffrey Bensusan to Peter Lord (22 August 1974), Box 100, CSDA.
52 Geoffrey Bensusan to Peter Lord (22 August 1974), Box 100, CSDA.
53 Ibid.
54 Clause 9.7, Code of Professional Conduct (1976), Box 100, CSDA.
55 Gordon Russell (1976), quoted in Fiona McCarthy, *Eye for Industry: Royal Designers on Design* (London: Royal Society of Arts, 1986), p. 1.
56 Raymond Spilman to William Renwick (11 February 1963), Spilman papers, Box 41, SCRC.
57 Walter Dorwin Teague to Peter Muller-Munk (16 June 1955), IDSA archives, Box 65, SCRC.
58 Jon Hauser to John Vassos (10 October 1963), Vassos papers, Box 1, Correspondence, SCRC.
59 Edgar Kaufmann, Jr to the IDSA Board of Directors (15 April 1965), Vassos papers, Box 2, Board of Directors, SCRC.
60 Ray Spilman, 'Redefinition of ASID Position and Objectives', Committee report (1963), Spilman papers, Box 41, SCRC.
61 Henry Dreyfuss to Raymond Spilman (9 July 1962), Spilman papers, Box 41, SCRC.
62 Ibid.
63 Ibid.
64 IDSA 7th Annual Symposium, 'The Professional Challenges of the 60s' (12 October 1960), Box 2, Board of Directors, Vassos papers, SCRC.
65 Henry Dreyfuss to John Vassos (28 July 1965), Vassos papers, Box 2, Board of Directors, SCRC.
66 Memo, (n.d., c.1968), Box 2, Board of Directors, Vassos papers, SCRC.
67 IDSA membership survey 1965, Box 2, Board of Directors, Vassos papers, SCRC.
68 Ibid.
69 Raymond Loewy, letter of resignation from the IDSA (18 February 1964), Ethics/Grievances folder, IDSA Archive, SCRC.

70 Brooks Stevens to Haggstrom (20 February 1964), IDSA Archive, 'Ethics / Grievances folder', SCRC.
71 Renwick to the Board of Directors (23 April 1964), Box 2, Board of Directors, IDSA Archive, SCRC.
72 Educational Seminar, Plaza Hotel, New York (15–16 April 1967), IDSA, Box 2, Board of Directors, Folder 6, Vassos papers, SCRC.
73 Memo (n.d., c.1968), Box 2, Board of Directors, Vassos papers, SCRC.
74 Ibid.
75 Memo, 'Climate of Discontent', F. E. Smith to the Board of Directors (20 May 1969), IDSA, Vassos papers, Box 2, Board of Directors, Folder 6, SCRC.
76 Ibid.
77 Arthur J. Pulos to the IDSA Board (26 May 1969), Vassos papers, Box 2, Board of Directors, Folder 6, SCRC.
78 Ray Spilman, 'Thoughts on Professionalism', originally published in *Oculus*, publication of the New York Chapter of the AIA (March 1959), Spilman papers, Box 41, SCRC.
79 See also Robert Gordon-Fogelson, 'Vertical and Visual Integration at the Container Corporation of America', *Journal of Design History*, 35:1 (2022), 70–85.
80 Spilman, 'Thoughts on Professionalism', Box 41, Spilman papers, SCRC.
81 Ibid.
82 Pilditch, Chapter 9, 'How to be Creative', in *The Silent Salesman*, pp. 98–112.
83 Ibid., p. 98.
84 Ken Garland, interview with the author (26 February 2013).
85 Ibid.
86 Ken Garland, *First Things First: A Manifesto* (London: Ken Garland, 1963).
87 Anthony Perks to *SIA Journal*, 140 (October 1964).
88 Lynton Lamb to Peter Lord (15 June 1968), Box 100, CSDA.
89 Ibid.
90 'Towards the Next Century: A Strategy for the Chartered Society of Designers', Box 13, CSDA.
91 William Goldsmith to John Vassos (27 April 1966), Vassos papers, Box 2, Board of Directors, SCRC.

6

Social responsibility and the industrial designer

The 1970s in many ways represented a new era for industrial design, as the profession pivoted towards a new audience: the public. A profession predicated on planned obsolescence now turned to face the consequences of its impact on environmental, social, health and ecological concerns. To describe this transformation as a reinvention is not to exaggerate the scale of change involved for the industrial designer's identity. While there were exceptions of course, the industrial design consultant had until now operated at a cultural distance from the public for which it designed and sold products. As Chapter 4 discussed, the professional ideal pursued by the professional organizations, including the SIA and IDSA, prioritized the designer–client relationship above all else. The public image of the designer, mediated through newspapers, magazines or on the exhibition stage, often took the form of an omniscient, 'all-seeing' expert; living a life of 'good taste' the general public could only aspire to. Even publicly funded organizations, like the CoID, sought to develop a relationship between design and the public that relied upon 'good taste' and standards of design, rather than social good, a fact for which it increasingly received criticism.[1] Professional guidance literature positioned the designer as an intermediary between the client and the public, serving as a 'middleman' between the two, but mostly aloft from public concerns.[2] The reinvention of industrial design as a socially orientated practice stretched the boundaries within which the profession had originally been established.

While, as Chapter 4 discussed, this vision of design had been mobilized on an international scale through the ICSID and its associated initiatives, professional organizations in the US and Britain moved more hesitantly and with some reluctance towards this reinvention. This chapter examines these attempts to respond to change and to make the professional ideal relevant in the final decade of the period under study in this book. It follows attempts within the SIAD and IDSA to formalize the designer's relationship to the public through the licensing of the industrial designer and to acquire a Royal Charter. The second section discusses the emergence of anti-professional critiques that formed the basis for Victor Papanek's bold statement that industrial design was the 'most dangerous profession in the world'. The chapter explores the struggle, still ongoing, to articulate a meaningful relationship between the industrial designer and social responsibility, and it ends by reflecting on the 'lessons unlearned' within the profession. In particular, the gendered identity of the male industrial design hero held strong in the field of socially responsible design.

'Human factors'

It is the responsibility of industry to produce well-designed and well-engineered products, not the responsibility of the consumer to demand them. Our programmes of planned obsolescence have set up an industrial complex geared to producing more waste and a society trained to accept it. Almost all of the best design and best quality is being produced in anti-people products – missiles, weapons, computers. Products for people – toasters, dishwashers, cars – are seldom designed to perform half as well.[3]

Speaking at an SID event 'Better Design for Better Business' in 1963, this powerfully direct statement by Hugh de Pree, President of the Herman Miller Furniture Company, identifies the shift from one 'industrial complex', the consumer society, to another, as design was being instrumentalized as a tool in cold-war politics. Having functioned to serve the interests of the corporation through planned obsolescence since its inception as a profession in the inter-war period, designers now faced a more complex and politicized relationship with the public. De Pree's division here between 'anti-people products' and 'products for people' reflected the line frequently drawn in contemporary critique between 'morally and ethically virtuous design' and design for 'profit-driven corporate culture'.[4] In reality, as design historian and social anthropologist Alison J. Clarke has argued, the lines between the two were increasingly blurred.[5]

By the middle of the decade, in the US President Nixon doubled funding to the National Endowment for the Arts and established the Federal Design Program, bringing professional organizations including the IDSA and AIGA into consultation for the first time and instituting a series of design assemblies 'charged with brokering the relationship between federal agencies and designers'.[6] The government turned directly to the 'pioneering' Consultant Designers to initiate this new relationship between the arts and government policy. In 1965, Raymond Loewy was invited to lead the US President's Committee for the Employment of the Handicapped, with Don McFarland leading the California team, Dave Chapman the Mid-Western team and Bill Raiser the East Coast team. The decision to appoint

Loewy to lead the initiative reflects the intention to bring a 'big name' to the project, in the absence of any specialized approach to the problem. Speaking of his role to the IDSA audience, Loewy said:

> This show of professional spirit in its purest form places the industrial designer on a plateau of idealistic abnegation that will leave its mark and add to the prestige of every one of us. Besides, for the first time in our history our entire profession is called upon to help our Government on solving a nation-wide problem ... On behalf of millions of Americans physically disabled, I thank you.[7]

This expression of public servility can be read in the context of increased engagement between the US government and the IDSA, as design was harnessed as a soft-power tool in 'cold-war cultural transactions'.[8] Ironically, a series of letters exchanged with the IDSA Board of Directors reveals concern about the unprofessional and unethical conduct of the government's contractual practices for this work, as they invited designers to contribute speculative work for the Atomic Energy Commission and NASA.[9]

In 1968, the IDSA annual conference was held in Washington, DC, with a VIP tour of the White House to kick things off.[10] The event started with an admission that the IDSA 'does not go far enough in claiming our social responsibility'.[11] The vagueness with which this term was used by the Society reflected the limited consideration that had hitherto been given to the industrial designer's relationship with the public, beyond the dynamics of production and consumption. Indeed, multiple interpretations and definitions of the 'social' were operating within the discipline and profession of industrial design internationally. This was shaped to a large extent by public discourse that had been, as Chapter 5 discussed, promoting the design profession's ability to tackle 'social' problems that included environmental catastrophe and irresponsibility; colonialism and neocolonial development, social injustice and global poverty. From within, the discipline of industrial design was also being reshaped by new 'user-centred' design methodologies, based on 'quasi-anthropological' scientific aims.[12] The Southern New England chapter IDSA Dinner Meeting in 1969 was on the topic of 'Human Factors'.[13] Here, the industrial designer was being turned to face its public, not as a consumer, but as a 'user', a differentiation that had been refined through new design research methods first developed in Sweden and Finland.[14] Industrial design had to move beyond obsolescence as its governing paradigm and find a way to address design for need.

While being awake to this myriad of complex disciplinary and political shifts in the integration of 'the social' in design education and practice, on an organizational level, the IDSA struggled to find ways to meaningfully articulate its role in relation to society at large, having been focused for so long on the question of its usefulness to the corporate client and, to a lesser extent, the consumer. This struggle can be read in the desperate exchanges between the IDSA Board of Directors in 1965, as they returned to this existential question when it set up a Licensing Committee to investigate ways of licensing the industrial designer as a professional in New York state, an initiative they believed would lead to a more secure professional status for designers in the entire country.[15] The committee circled back to questions of licensing that had been pursued by the SID and ADI independently (and competitively) in their founding years.[16] However, the IDSA's obscure relationship to 'the public' remained an obstacle to this. As Henry Dreyfuss wrote to John Vassos in August 1965, 'in order to acquire licensing in New York state, you have to be involved in a profession that has to do with the safety of the public'.[17] A legal consultant, Sam Roman, was employed to investigate the issue.[18] According to notes prepared for the (unsubmitted) application, the 'handicap program' was to form the backbone of the IDSA's claim to social responsibility.[19] Roman recommended a PR writer in industry to prepare a 'clear and definitive document outlining the profession of industrial design, its structure, its breadth, its application of special knowledge of materials, its involvement with safety factors, its support of students through awards, its relationship with industry etc.'[20] Arthur J. Pulos wrote with some concern about the IDSA's approach to the subject, stating, 'it worries me that a PR writer in industry turns out to be the best qualified person available to write a clear and definitive document on industrial design. It all sounds so commercial and non-professional.'[21] The fact that the other Board members did not identify a conflict of interest here further reveals the close relationship between publicity and professionalism for the IDSA and its founding members, for whom professionalism was utilized as a promotional tool in leveraging greater status and economic security.

The IDSA report on licensing sought to illuminate 'designing with the welfare and protection of the consumer in mind', 'to secure the profession from the inroads of the unqualified' and to 'enhance the stature of the field for those who enter after four or five years of specialized study'.[22] It was noted that in 1958, the American Institute of Interior Designers attempted to obtain licensing in the State of California, but had been unsuccessful because of a 'failure to clearly establish the public good as the objective'.[23] As he explained, 'licensing would be positive proof of the good intent of our small professional group, not only in the automotive industry but in every area of endeavour where the protection of human beings is a vital factor ... it is our job to protect the future of the professional by providing designers with a sense of obligation to and awareness of their public responsibility'.[24] This was repeated many times by many different people on the IDSA Board, but no one seems to have had a clear idea about how to do it. As Julian Everett, chairman of the IDSA Licensing Committee, wrote to Vassos in August 1965, 'under the present conditions, we are bumping our heads against a stone wall of entrenched privilege and official disinterest'.[25]

Indeed, the correspondence on the issue of licensing presents the IDSA in a state of paralysis on questions of self-definition and identity.

The report used capital letters: 'WE MUST ERASE THE IMAGE OF THE STYLIST and the way open to us to do this is through Licensing', a statement that revealed how little the Society had moved on from questions addressed in formative meetings of the SID in 1945 discussed in Chapter 1. Seemingly forgetting the Society's professional ethics, Henry Dreyfuss rather desperately proffered the option of resurrecting a role for designers in corporate advertisements, as a means of calling attention to the profession: 'I always thought it was very complimentary and flattering to have the client use the designer's name in advertising and in publicity releases. Do you think we could reawaken this somehow?'[26] In 1965, Arthur J. Pulos wrote to William Katavolos to say that the profession's 'state of mind must go through drastic change' before industry and the profession would subscribe to licensing.[27] By 1968, the discussion of licensing was effectively dropped from the priorities as there was 'no clear overall mandate from the membership to proceed in any particular direction'.[28] The tone of the IDSA discussion was disengaged and resigned – it seemed that no one involved had the appetite or the ability to take on this issue.

As the previous chapter described, the dominant position of the corporate designer within the membership contributed to a more practical set of concerns and debates. As one member eloquently, but critically, put it:

> IDSA is moderately effective and certainly worthwhile; it can be wholly effective only when we reach an agreement on what Industrial Design is ... If we can agree that industrial design is the profession most responsible for the broad scope of human-to-product interface including a moral conscience responsibility for the intrinsic value, safety and esthetics, then we stand a chance with the help of IDSA of finding the security and permanence afforded by filling a genuine social need.[29]

As this anonymous member seemed to intimate, indications had so far suggested that the IDSA was struggling to face the challenge of fulfilling this new social role for design. In 1970, the IDSA called upon the consultation of publicist Betty Reese, who 'cautioned against the mouthing of backing for public service projects unless we are really involved – we should not over-estimate the dedication of our members'. Presumably informed by her many years of service in the Loewy office, Reese made the point that, in fact, 'it might not be realistic to expect designers to be socially conscious'.[30] She reminded the Society that they still needed to identify places of public interest – environment, behaviour, government, education. It was a daunting task and one the IDSA was clearly not fit to address from within.

Royal Charter

Responding to similar critiques and debates, the Society of Industrial Artists and Designers (SIAD) in Britain also sought to formalise the designer's relationship with the public in the context of social responsibility. One way of doing this was by establishing a more direct association with education and learning. During the 1950s, a formal system for student membership, Licentiate membership, was established. This was awarded to students from an SIAD-approved course in design – usually those registered as National Diploma in Design (NDD).[31] After the establishment of the National Council for Diplomas in Design (NCDAD) in 1961, the Diploma in Design (DipAD) courses were automatically validated by the SIAD. This was known as the Direct Admission Scheme and became a 'pipeline of SIAD membership' throughout the 1960s, drawing in a geographically diverse cohort of student members from courses across the country. The Society's membership grew substantially as a result of this scheme.[32] However, it also intensified generational tensions within membership, as student members voiced their critique of the Society's aged and outdated gentlemanly view of professionalism.

The SIAD made a further claim to professional legitimacy in 1976, through the acquisition of a Royal Charter, an 'inter-association status symbol, a distinguishing mark, acknowledging supremacy in a particular field and the ability to provide a sound public service'.[33] The Royal Charter 'constituted the final break with the original passive notion of the Society as a club. Instead, it was now conceived as an independent, private, proactive body, with public responsibilities.'[34] The acquisition of a Royal Charter was one of founder Milner Gray's longest and most abiding aims for the Society, but it had been rejected on two previous attempts. Prime Minister Margaret Thatcher, an advocate of sorts for the industrial design profession in Britain, played a role in securing Royal Assent for the Society.[35] It was not the first professional design organization to be awarded this status.[36] Ironically, it was also awarded at a time when the Society's reputation was in decline, following the 'professional crisis' of the late 1960s, and membership figures were dropping. Many students who automatically entered the Society's membership through the Licentiate Scheme did not renew or apply for full membership status upon graduation. Announcing the transition, President Richard Negus used the opportunity to symbolize the Society's new-found connection to public service:

> Now that the Charter exists and we have the standing that was essential … a case can be made for spending less time on such things as rules of behaviour and their interpretation and concentrating on reducing the gap between the designer and the public. The title 'designer' now has a meaning that was absent 25 years ago. Each designer bears a major responsibility to attain and demonstrate the highest standards … Good design and membership must become inseparable in the public mind. So, welcome back Terence Conran![37]

For those within the Society, Conran represented a commercial approach and was proof that it was more profitable to work beyond the narrow conception of the professional ideal to which the Society had been previously bound. The decision to readmit him symbolized a determination to move beyond the values that had previously held the Society so tightly together. Conran was awarded the SIAD medal in 1980. He wrote to President George Freeman to accept with surprise, and Freeman replied, 'I do not think you should be surprised by the medal. The Society has a great deal more respect for you and your work than you evidently realise.'[38]

In 1978, a 'Forward Planning Committee' was appointed to make predictions about the future of design. At one of its first meetings, Eddie Pond eloquently put it, 'There are too many old fogies and pompous bastards involved and we must get some young people doing things.'[39] In an article in the *Designers Journal* (1984), interior designer Stefan Zachary argued that the recent acquisition of the Royal Charter marked a 'final break away from the corduroy jacket and roll-your-own brigade'.[40] That year, the Society again seized upon a narrative of change, and commissioned James Pilditch to write a report on professionalism entitled 'Towards the Next Century: A Strategy for the Chartered Society of Designers', in which he made an even bolder assessment on the limits of professionalism in design: 'There was a time when the Society tried to fix fee scales, forbid advertising and prevent competitiveness. All were thought intrinsic to professionalism. None lasted.' It was a desperate attempt to appeal to a contingent beyond the limits of the professional ideal. Membership numbers continued to decline. Pilditch, a sharp businessman, noted the 'dynamic' business environment in which designers now practised in Britain, with the more focused attitude of the Design Council and the imminent opening of a new Design Museum in London. 'The grown man is not the same as the boy', he said, plainly restating the masculine identity of the designer. 'Design has grown and so has it evolved.' More broadly, he argued, 'professionalism is undergoing change. Long established privileges are eroding. Anything that fosters restrictive practice is being blown open', a statement that points towards the alignment of design with neoliberal market practices.[41]

By 1980 the CSD shared premises with the D&AD in Nash House, a project to unite organizations of similar aims and objectives in the auspicious residence of Carlton House Terrace, west London. This residence of 'crumbling stucco grandeur', as one design critic put it, was an apt setting for a society of gentleman-professionals and the fading relevance of the professional ideal in British design more generally.[42] An article in *The Designers Journal* in 1984 entitled 'To Join or not to Join?' weighed up the benefits, 'if any', of joining a professional body.[43] A photograph of the Society in the article reveals the ageing image of the CSD and its failure to include women at the upper levels of its organization. The only woman in the picture, June Fraser, was finally appointed President in 1983. As Robert Wetmore, President of the SIAD in 1976, revealed in an interview with Fraser a few years later, the Society had been desperately trying to find ways to get 'ladies' to do things.[44]

In an effort to identify and reach across to new audiences and publics, professional organizations in Britain and the US now sought to bring critics and 'outsiders' of the profession back into its sphere of influence, with limited effect. As discussed in the previous chapter, in 1964, Raymond Loewy furiously tendered his resignation to the SID, on the basis of a grievance between himself and Brooks Stevens.[45] The SID Board was concerned that the resignation would affect the planned merger of the ASID and IDI. Loewy stated that this was 'far-fetched', but 'I could accept some form of Honorary Membership.' The indifference of this remark suggested the limited value and relevance the SID held for him, a feeling shared with his co-founder Walter Dorwin Teague. Seeking to reach out to new constituents in August 1967, the IDSA proposed an invitation to membership by Walter Landor, a designer previously considered too corporate and commercial, with full endorsement from the San Francisco chapter.[46]

A dangerous profession?

Seizing upon the reduced agency of the professional organization, designers including Victor Papanek played with the dissolving boundaries of professionalism, turning them to personal advantage. As Clarke has argued, Papanek had been driven to 'counter the industrial rationalism of his modernist forebears' as early as 1946, envisaging a human-centred mode of design, but these ideas had finally reached an audience within the profession itself by the 1960s.[47] While Papanek was not the first industrial designer to present the pressing need for designers to turn to social responsibility, his book *Design for the Real World* helped to popularize the movement. The famous opening paragraph of the book presented perhaps the most dramatic portrayal of the profession's destructive capacities yet:

> There are professions more harmful than industrial design, but only a few of them. Never before in history have grown men sat down and seriously designed electric hairbrushes, rhinestone-coloured boxes, and mink carpeting for bedrooms, and then drawn up elaborate plans to make and sell these gadgets to millions of people. Today industrial design has put murder on a mass production basis. By designing criminally unsafe automobiles that kill and or maim nearly one million people around the world each year, by creating whole new species of permanent garbage to clutter up the landscape, and by choosing materials and processes that pollute the air we breathe, designers have become a dangerous breed.[48]

Papanek's polemic built upon the critiques of the 'new professions' by C. Wright Mills, Vance Packard and others described in the previous chapter. Nevertheless, in focusing squarely on the industrial design profession, its structures, the designer's reliance on corporation and the culture of planned obsolescence, his critique dismantled almost everything the profession had been built upon. Moreover, his rhetoric playfully mocked the egos of the 'grown-men' Consultant Designers who had dominated the image of the profession, questioning their aspirational status as taste leaders they had worked hard to construct. At a time when professional organizations, including the IDSA, were desperately searching for a way to articulate their relationship to the 'public' and their commitment to 'social good', Papanek made it crystal clear that these goals were incompatible with the values and ideals of professionalization. It put in stark terms how far the profession would have to move to achieve these goals.

Papanek resigned dramatically from the IDSA in 1969, citing frustrations with its management,[49] although it later transpired that he felt particularly aggrieved by a meeting at the April 1969 Annual Conference.[50] Much to his further frustration, his letter of resignation was ignored (apparently 'lost' or 'misfiled' by the IDSA secretaries),[51] so he wrote to the Society again a year later to confirm his resignation. Raymond Spilman, who was actively involved in the IDSA Board of Directors and would become President in 1972, wrote to Papanek to placate him, urging him to reconsider. 'No-one could top a designer in managing to feel unwanted, ignored, hurt, otherwise insulted ... We designers tend to be bitchy types.'[52] Moreover, Spilman recognizes Papanek's value to the IDSA as a designer who had identified the value of 'the social' in his work, stating:

> I believe we need you as the practising teacher and professional that you are ... I do think the Society has moved toward maturity and greater social responsibility and hopefully it will continue in this direction. People like you Victor, in fact, you have made great contributions to our perhaps to [sic] slow but growing awareness of what the responsibility of industrial design is all about. Don't go away or be defeated by the timid, inconsiderate or rude – the majority need you more.[53]

Spilman's wording here indicates the symbolic role Papanek held in stimulating critical conversations about professional responsibility and design in the US, where, as we saw in Chapter 4, forums for this kind of dialogue had previously been limited and constrained by the corporate agenda. Perhaps buoyed by the recent popular success of his book, Papanek wrote to Goldsmith on 23 June 1971:

> John [Vassos] was, as is usual, right: when I wrote my original letter to you, I was under the impression that I was staying. As it is, I consider it unacceptable that in less than three weeks, when I attend the ICOGRADA/ICSID meeting in Vienna, I shall have to rely on my honorary memberships in three Nordic industrial design societies to be permitted to attend.[54]

In making his support of John Vassos clear, Papanek was hinting towards his alignment with the IDI, previous to its merger with the IDSA. As discussed in Chapter 1, the IDI was known to have practised more open and inclusive attitudes to professionalism in design, representing a wider geographical, gender and disciplinary basis for its membership than the New York-based, consultant-led SID.[55] Vassos, founding member of the IDI, was apparently a supporter of Papanek.[56] Clearly enjoying the new-found status the popular success his book had given him, he wrote to Goldsmith:

> For your information, I have instructed my publishers that the initials IDSA should not appear after my name. On the jacket of the Swedish edition, they do not appear. They will also not appear on the American, Canadian, British, Danish, German, French, Finnish, Czechoslovakian or Spanish-language editions. However, I have so far been unable to get through to my Italian publishers, and I am sorry about that.[57]

The dramatic tone of Papanek's rejection of the IDSA accreditation and its professional values makes it difficult to understand the fact that within less than a month, he had reapplied to the Society for membership. Replying to this reapplication on 14 July 1971, Goldsmith wrote:

> I find it difficult to read into a number of your observations and the way you say them any sign of affection or respect for IDSA and most of its members. I am, therefore, puzzled as to your interest in rejoining. I can only assume that you are expressing some frustration at our contacts in the past 20 months and that you are, really, interested in contributing to our common purpose. I do hope that is the case and that you are desirous of putting your unquestioned talents into constructive channels rather than persisting in such a large dose of criticism.[58]

While Goldsmith assured him that there would be nothing to prevent him rejoining, it would go to the Board of Directors and the decision would rest with his peers, 'as it should be'. 'Let bygones be bygones', he said.[59] Papanek, a keen agitator of the profession, possibly concocted the dramatic resignation in order to confirm his outsider status at a time when his reputation depended on it. According to Clarke, Papanek later claimed that he was 'blackballed' by the IDSA and was denied membership in 1964, a statement that does not fit the evidence presented in this correspondence.[60]

In fact, Papanek's professional alliance had been with the IDI, of which he was a member for nine years. In 1964, Papanek had written to John Vassos at the IDI to ask for a reference to support his application to the Kaufmann International Design Awards, administered by the Institute of International Education. These awards, set up in 1937, were sponsored by the Edgar Kaufmann Foundation 'to give public recognition to accomplishments in all fields of design and to encourage enterprising design developments'.[61] Papanek wrote to Vassos that he had been 'invited to apply' by Robert Malone, editor in chief of *Industrial Design*, who was consultant to the awards programme and aware of Papanek's 'keen interest in design for the underdeveloped and backward peoples of the world'; language that revealed the cultures of Western superiority driving design for development discourse.[62] His proposal articulated the new political and social currency of design:

> As a profession, we are designing and teaching young people to design for a mythical, middle-class, middle-income family in the middle West with all the power resources at their command and at the available annual income of $2,800 annually. Since the mean annual income now stands at $118 and the power-slave units at the command of the individual vary from 585 per person in the US to 1/18th per person in certain areas of South East Asia, it is time that the industrial design profession, master form givers to all mass-produced things in this world (with the possible exception of housing and apparel), move ahead into areas of meaning. The needs of the people of the world in areas of shelter, transportation, care for the sick, the retarded, the handicapped, communicator from government to people and from people to people etc., are the challenges that must be met.[63]

Papanek was clear-minded and pragmatic in giving his justifications to the committee on why they should award his research: '1) Just plain good business to satisfy the wants of the two billion, 350 million people who live in need. 2) The US image abroad would be helped fantastically. 3) The Soviet Union and some of the countries behind the iron curtain are now working in these directions and if Mr Khrushchev's boast to "burry us" is to be taken seriously, it is in this area that the burial will take place.'[64] Vassos was supportive of the application, although it was ultimately unsuccessful and the awards themselves were wound down to a close by 1967. Nevertheless, the language of Papanek's application reveals the pragmatism of his approach, as he presented his research aims within the rationale of promoting industrial design 'in the periphery' as a containment of communist expansion.[65] In this sense, he shared the pragmatism of the pioneering designers he so ardently critiqued, Raymond Loewy and Walter Dorwin Teague.

The design team

Papanek's 'The minimal Design Team' infographic, originally produced as part of the Big Character Poster No. 1: Work Chart for Designers (1969), has become a visual touchpoint for historians in pinpointing new directions for the industrial designer's professional configuration (see Figure 6.1). Circling back to the question of specialist versus generalist expertise that had dominated design discourse in its formative years in the US, Papanek's vision for the 'socially responsible designer' now imagined the designer as a generalist, a 'co-ordinator of specialists', in a language that was resonant of that of Henry Dreyfuss and the 'pioneers' of US industrial design. Once again, the designer was the 'middleman', an 'interpreter' between disciplines. However, whereas in the mid-twentieth century,

'The Industrialized designer'

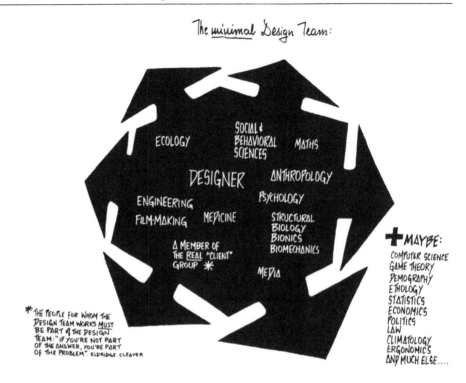

6.1 'The minimal Design Team', detail from Big Character Poster No. 1: Work Chart for Designers (1968 [1973]). Courtesy of Victor J. Papanek Foundation, University of Applied Arts Vienna.

this had involved the co-ordination of design within the corporation and its attendant management structures, here Papanek was articulating the designer's expertise in relation to the social, behavioural and environmental sciences. In stating this to be the 'real' client group, he was dismissing the long-standing centrality of the corporation on the industrial designer's practice and embracing the wider 'anthropological turn' in design.[66] Importantly, Papanek's configuration of the Design Team put the 'user', a non-professional, at the design table. As such, his vision for the industrial designer's role reached out, as indicated in the outward-pointing arrows, beyond both the profession and the academy and into the public.

Papanek's writing raised the significance of the 'design team' in professional design discourse in the US, importing the values of experimental co-design methodologies that had been first developed in Scandinavia.[67] However, as the first two chapters of this book argued, the design 'team' or 'unit' formed a central motif within professional design discourse in Britain from the inter-war years through to the 1960s Design Methods movement, where the anonymity of the designer was secured through a firm commitment to the idea of design work as the outcome of a group of specialists.[68] This was intrinsic to the organization, structure and identity of the first British design consultancy, the Design Research Unit (DRU), which had, since its institution, worked to the principle of flat organizational structure, with teamwork at its core.[69] The design team was also central to the SIAD's ideas of professional ethics. Architect and industrial designer Jack Howe, President of the SIA 1963–1964, confirmed this in his 1963 address to the Society, entitled 'Social Responsibility'. Whereas the craft worker 'works to please himself and the solitary agent', the designer, Howe stated, 'cannot opt out of this responsibility':

> The social responsibility of the designer involves him not just with people and social attitudes at a theoretical level. It involves him with other people who are also highly trained essential members of the design team – the economist, the sociologist, the scientist, the engineer and the manager.[70]

Howe's emphasis on the 'highly trained' expertise of the designer reveals the fundamentally professional, scientific view of design work referenced the approach of the Systematic Design Methods movement.[71] His view of specialized, professional expertise was fundamentally at odds with Papanek's, which put the non-professional 'user' at the centre of the team. While *Papanek's Design for the Real World* was not published in Britain until 1971, he was warmly received there and took up a position as principal lecturer in the Manchester Polytechnic design school in 1973, the same year that the 1968 Copenhagen Flowchart, in which 'The minimal Design Team' was drawn, was made into a full-sized wallchart.[72] In 1973, he delivered a paper at the SIAD conference, 'Design for Quality in Life', and the ICSID conference of the same year was entitled 'The Relevance of Industrial Design', events which Penny Sparke argues were responding to Papanek's critique, 'channelling an already latent crisis of conscience into tangible form'.[73]

Design for development

Under cold-war conditions, industrial designers practised in a new sphere of political influence.[74] In the 1970s, the ICSID turned to international organizations and agencies to pursue its aims, gaining consultative status with the United Nations Industrial Development Organization (UNIDO) in 1974.[75] In 1976, it signed a joint memorandum with UNIDO to 'jointly accelerate industrial design activities', which developed into the 'UNIDO/ICSID' 1979 Conference in India. Based on what Clarke terms 'quasi-anthropological research paradigms', a specific 'design for development' agenda emerged, in which the industrial designer would play a key role as a mediator between soft-power cultural diplomacy and industrial development programmes with economic expansionist aims. Here again, the designer's status as an intermediary, or 'middleman', was harnessed for corporate agendas and soft-power strategies in the context of aid and development programmes. Beyond the remit of obsolescence, designers found themselves in uncharted territory, with very little in their professional or educational equipment to guide them. In spite of the rhetoric and polemic, the fact that designers were rarely trained, nor professionally equipped to work in these areas soon became apparent, as examples of misguided, opportunistic and even exploitative work by Western designers in developing countries emerged. Arguably, this new remit for the profession had potentially more 'dangerous' implications than the one Victor Papanek had critiqued. Shocking accusations of corruption and exploitation by individuals working under the guise of 'design for development' were sent to professional organizations, including the IDSA, filed in the 'grievances' folder, but no action was taken. It was clear that this behaviour now fell outside the jurisdiction of the national professional organization and its ethical Code of Conduct.

In 1970, the ICSID formed a 'Working Group Developing Countries', which included Nathan Shapira in Nairobi, Kenya and Gui Bonsiepe in Buenos Aires, Argentina, as co-ordinators. Shapira drafted the first UNIDO 'Design for Development' strategy report, but Bonsiepe, who was instructed by UNIDO to write the first report, stated in an interview that it had 'substantial shortcomings', 'mainly because this colleague' (referring to Shapira) 'didn't have substantial first-hand experiences in a developing country context'.[76] Two years later, the Design Society of Kenya sent a shocking, angry letter to the IDSA to make an official complaint against the work of Shapira, 'Visiting Professor' at the University of Nairobi, who had been there for three years with the task of advising the University and its staff on the academic objectives for Design Education in East Africa. The emotionally charged frustration of the letter may have been further enhanced by the fact that, as Daniel Magaziner writes, the University had previously been engaged in more sensitive, postcolonial approaches to industrial design as a way of 'reaching into the future'.[77] The authors of the letter provided a considerable list of allegations against Shapira, including 'selfish aims and financial hoarding', stating that 'he was at no time concerned with the real needs of our people'. They claimed that Shapira had used 'departmental funds to produce and distribute reports on the ICSID Ibiza meeting on Developing Countries, for distribution only to the members of the Board of ICSID, and members of the Committee. No copies were distributed to Kenyan designers or design educators.' Shapira was, they said, a 'dangerous designer; dangerous and harmful to the dreams of people in developing countries; dangerous to the future of Design Education in whatever school or country he might choose to go next' and responsible for the closure of the 'first and only department of design in Africa'.[78] They urged the IDSA Board to support their objections to 'his representation in ICSID and his official status at the Ibiza Congress, which was not in any way supported by us designers from East Africa'. Replying to John Vassos in reference to the allegations, designer and IDSA President Tucker Madwick makes no recommendation for further action, glibly comparing his work to Victor Papanek's, stating, 'maybe one day their paths will cross', making clear his shady view of work in the design for development field and his reluctance to become involved.[79] Shapira's name is associated with a series of questions and concerns throughout the 1960s, but IDSA Board members seemed unmoved, or unable, to take any firm action against him.[80]

Lessons unlearned

While the field of industrial design had arguably never been bigger or more ambitious as its agency was dispersed to wider and broader constituencies than ever before, the 'first generation' of the industrial design profession continued to present itself publicly in ways that were remarkably unchanged from the inter-war years. In 1970, Arthur J. Pulos published *Opportunities in Industrial Design*. In his foreword to the book, Tucker Madwick, President of the IDSA that year, wrote:

> Many tomes have been written about the professional practitioners of industrial design; such as Teague, Loewy, Dreyfuss, Deskey, Bel Geddes. But how does one become a professional? To date, very little has been expounded to aid the uninitiated in his search for a career in industrial design. The purpose of this manual is just that, to help the reader achieve a career in the design field, what subjects to explore, what schools to attend, whether corporate or consultant.[81]

In its language, the book continued to refer to the professional identity of the industrial designer as male and continued to divide the role between corporate and consultant work. Pulos again referred to industrial design as 'one of the new "generalist" professions, which will be vital to our society in the next century' in language that harked back to the founding of the profession, stating, 'Industrial designers define the character of a company to the public on one hand and describe public demands and wishes to the company on the other. They serve as an indispensable link between people and their industries.'[82] In spite of the turbulent changes in the intervening thirty years, the profession he describes was in many ways unchanged in its essential outlook and function from that expressed in Harold Van Doren's 1940 *Industrial Design*, described in Chapter 1. Just as Van Doren had expressed difficulty in defining the role of the industrial designer, Pulos conceded that there was 'difficulty in defining the role' because the profession is 'young and alive' and 'it would seem that the industrial designer aspires to be many things to many people'.[83] This blurry admission reflected the very limited progress the IDSA had made towards clear professional status or identity.

'The Industrialized designer'

For those in the CSD in Britain, the central tenets of professionalism, inherited from the traditional professions, stood strong. In her third and final edition of *The Professional Practice of Design* (1978), DRU manager Dorothy Goslett stated:

> Many changes in the seven years since the previous edition of this book was written have made revisions necessary ... the most important were the significant changes made by the SIAD to its Code of Conduct whereby designers were given much greater freedom their services, even to the point of advertising them.[84]

In actual fact, Goslett's revisions were relatively minor, as she sought to defend the SIAD's outlook on professionalism. Nevertheless, the book now included a definition of the concept of 'promotion', a new word in the SIAD's professional vocabulary ('to publicise and sell in all appropriate and permitted ways and includes advertising, editorial publicity, writing direct to potential clients, appearing in public and on the air and eventually paying experts to do this for you').[85] Goslett couldn't resist interjecting here to defend the gentlemanly ideal of professionalism:

> Contrary to what the gloomsters predicted when clause nine became an accepted code, the design profession has had the good sense not to rush jubilantly into the media with full-page advertisements in the *Times* and the colour supplements and slots on commercial TV. Apart from the fact that they probably couldn't afford it, established designers have realized that to be discreet in their approaches is far more helpful to those young designers who are only just beginning to climb the ladder.[86]

In this way, Goslett attempted to render the professional ideal compatible with the professional tradition. The book made little concession to social responsibility or to the industrial designer's relationship with the public.

Those in positions of power within the IDSA struggled to move beyond the mentality of designing for planned obsolescence. This became apparent during the planning for the 1974 National Endowment of the Arts Grant to produce an exhibition, film and publication on 200 years of American design. The Bicentennial project represented a significant step towards greater recognition of design in the mechanisms of government. A committee of eleven professional organizations collaborated on the project, sharing an office and meeting to coordinate an exhibit that would represent 200 years of American design to the public.[87] The suggestions put forward focused on consumer products – from Levi's jeans to the Coke bottle and the 'hard hat'.

George Nelson, ever an astute observer of the profession, wrote to William Goldsmith, President of the Bicentennial Project, expressing grave concerns about the project plan: 'My fear is that the way it is sketched out would lead to a trap and the result of our focusing on what great fellows we are to the exclusion of real concern for "public awareness" ... I cannot believe that any listing of the great American design contributions, whether in film or exhibit, is going to turn any audience on.'[88] Instead, he advocated a more general introduction to design as a basic human activity: 'The designer, in other words, has moved from being an ordinary average citizen to an increasingly specialized performer who is now moving into generalist-type activities and all of these are responses to a changing social context which affects everyone.'[89] In addition to addressing the social responsibility of the designer, Nelson was also addressing the function of industrial design as a service industry, recognising the definitions and goals of industrial design that had been articulated within the ICSID by Tomás Maldonado and Josine des Cressonnières. Designers, Nelson argued, should not be a 'servant of technology', but a 'guardian of the natural (and synthetic) environment':

> We are all now aware of water and air pollution, thermo-pollution, the dangers of upsetting the ecological system, the approaching end of planetary resources which can be mined and the approaching beginning of a planetary accounting system which will be based on our annual income from the sun because that will be all we will have. The current shifting of the designer's main activity from that of a servant of technology to a concerned guardian of the natural (and synthetic) environment is a direct result of the pressures now being brought to bear ... the designer in other words, has moved from being an ordinary average citizen to an increasingly specialized performer who is now moving into generalist-type activities and all of these are responses to a changing social context which affects everyone.[90]

The project seems to have prompted Nelson to a stark realisation of the contrast between the destructive capacities of design, as seen from the outside, and the superficial, corporate vision of design offered by its professional organizations. It was not a lesson the IDSA was able to take forward, as it continued to pursue the project with goals unchanged.

Conclusion

Writing in *Industrial Design* in 1973, John Vassos described the passing of Henry Dreyfuss as the 'end of an era' in US industrial design, an era of 'strong personalities of individualists and professional pioneers'.[91] Nevertheless, it did not take long for a new narrative of pioneers to emerge, following the pattern laid down by the generation before them. Papanek, Fuller, Bonsiepe, Maldonado and others gained their own 'cult' following as 'giants' of the movement or, as Clarke puts it, 'the romanticized pioneers of a socially inculcated vision of design'.[92] The narrative had been changed, but in the West, it was still largely controlled by a group of white, male design consultants.[93] Nathan Shapira would go on to be revered as a 'pioneer' of socially responsible design and an archive was set up in his name at the San Francisco State University.[94] Victor Papanek may have been crudely referred to as the 'Pioneer of Patronising Design' by a British design critic in 1985, but the politics of his work has only recently started to be critiqued by design historians.[95] The structure and organization of the profession had changed significantly, but the cultural tropes associated with the identity of the Consultant Designer held strong. For all its dramatic posturing, it is difficult to identify a 'turning point' in relation to questions of gender or race in a profession that had been 'invented' by and for white, privileged men. This is a somewhat surprising and uncomfortable realization for a field engaged in questions of 'the social'.

The dissolution of professionalism as an ideal for industrial designers in Britain and the US was unmistakable. In 1976, ICSID hosted the 'Design for Need: The Social Contribution of Design' conference at the Royal College of Arts (RCA) in London. Topics addressed included 'the role of the industrial designer in disaster relief' and 'industrial design in dependent countries'.[96] As Lilián Sánchez Moreno has argued, the conference represented a significant moment for Western countries to adopt 'social responsibility as a mechanism of professionalization', to 'gain legitimation on behalf of the wider circle of recognized professions'.[97] Nevertheless, given the evidence outlined in this chapter, it is difficult to see this effort as a great success. As design historian Pauline Madge noted in her review of the literature that accompanied the 'ecodesign' movement, disappointingly, a relatively 'one-dimensional picture' of green design emerged.[98] Too often, she noted, the literature got caught up in the web of its own 'anti-design critique', a description that also summarizes the struggle of the design profession towards social responsibility. The following year, Arthur J. Pulos and Misha Black hosted a joint seminar hosted by the RCA, London with his industrial design students from Syracuse, on 'British/American Industrial Design'. Questions on the agenda included ' the challenges of altruistic design in the "real world"' and 'can present trends in the world economy support a future market of consumer products?' While they had followed parallel paths of professionalization, they now faced the same issues and both self-identified as in a state of professional crisis. They turned to face the questions, but as founders of the profession in both countries, they were both somewhat constrained in their ability to guide their students to viable solutions. In a provocative letter, Professor and Chair of Industrial Design at Auburn University, Alabama, responded to Arthur J. Pulos, who had criticized a publication drafted by his students, questioning the relevance of the IDSA. He asked Pulos, 'are you afraid to answer it?'[99]

Notes

1. Eleanor Herring describes this critique in *Street Furniture Design: Contesting Modernism in Post-War Britain* (London: Bloomsbury, 2016), pp. 164–6.
2. For example, 'How To Use an Industrial Designer' (15 April 1958) stated: '[The industrial designer's] usefulness to his client is in part a result of his fresh, independent, outside viewpoint, in part a result of the diversity of his experience. He represents your public', IDSA62_16_58–59, SCRC.
3. Hugh de Pree, President of Herman Miller Furniture, Design for Better Business, Art and Business Worlds Unite for Product and Environmental Design Improvement (15–17 May 1963), IDSA Archives, Box 1, SCRC.
4. Alison J. Clarke, 'Design for the Real World, Contesting the Origins of Social Design', in Claudia Mareis and Nina Paim, *Design Struggles* (Amsterdam: Valiz, 2022), p. 89.
5. Ibid.
6. These took place in 1973, 1974, 1975 and 1978. See Jonathan Woodham, 'Formulating National Design Policies in the United States: Recycling "The Emperor's New Clothes"?', *Design Issues*, 26:2 (Spring 2010), 27–46.
7. Raymond Loewy, Presidents Committee for the Handicapped, led by Loewy, IDSA Annual Meeting, (2 October 1965). IDSA Archive, Box 2, Board of Directors, SCRC.
8. See Harriet Atkinson and Verity Clarkson (eds), Special Issue: 'Design as an Object of Diplomacy Post 1945', *Design and Culture*, 9:2 (2017).
9. Leon Gordon Miller to the IDI Board of Directors (3 July 1962). Raymond Spilman to Dave Chapman (26 February 1964). IDSA Archive, Box 2, Board of Directors, SCRC.
10. 1969 Program Prototype, 1969 IDSA Annual Meeting (9 October 1968), IDSA Archive, Box 64, Human Engineering, SCRC.
11. Ibid.
12. Ida Kamilla Lie, '"Make Us More Useful to Society!": The Scandinavian Design Students' Organization (SDO) and Socially Responsible Design, 1967–1973', *Design and Culture*, 8:3, 327–61. See also Alison J. Clarke, '"Actions Speak Louder": Victor Papanek and the Legacy of Design Activism', *Design and Culture*, 5:2 (2013), 151–68.
13. 'Human Factors', the Southern New England chapter IDSA Dinner Meeting (1969), John Vassos papers, Box 2, Regional, SCRC.
14. Maria Göransdotter, 'Designing Together, On Histories of Scandinavian User-Centered Design', in *Nordic Design Cultures in Transformation* (London: Routledge, 2022), pp. 157–77.
15. Licensing refers to the authority given by a state or institution to perform a specified service and is a common feature of professionalization. See Everett C. Hughes, 'Professions', *Daedalus*, 92: 4 (1963), 655–68.
16. As Vassos explained to Henry Dreyfuss in a letter on 3 September 1965, 'Alexander Kostellow, Belle Kogan, Ann Franke and I went to Albany for a preliminary hearing on the subject. The results was that we were advised that both Societies (ADI/ASID) sponsor such a request and ASID declined. So we dropped it.' John Vassos to Henry Dreyfuss (3 September 1965). John Vassos papers, Box 1, Correspondence, SCRC.
17. Henry Dreyfuss to John Vassos (20 August 1965). Vassos Papers, Box 1, Correspondence, SCRC.
18. Henry Dreyfuss to Bill Katavolos (10 September 1965). Vassos papers, Box 1, Correspondence, SCRC.
19. 'Licensing for Industrial Designers' report for IDSA (13 May 1966), Vassos papers, Box 1 Correspondence, SCRC.
20. Memo to the IDSA Board of Directors from John Vassos, FIDSA (3 March 1966), IDSA Archive, Board of Directors, SCRC.

21 Arthur J. Pulos to Ramah Larisch (10 March 1966), Box 1, Correspondence, Vassos papers, SCRC.
22 'Licensing for Industrial Designers' (IDSA, 13 May 1966), Folder 6, IDSA Records, SCRC.
23 This initiative was led by Miss Beulah Spiers: 'the failure of the effort has been attributed to over-complication of the case and a failure to clearly establish the public good as the objective'. 'Licensing for Industrial Designers' report.
24 Arthur J. Pulos to the IDSA Board (26 May 1969), Vassos papers, Box 1, Correspondence, SCRC.
25 Julian Everett, Chairman, Committee on Licensing, report to IDSA Board, August 1965, Vassos papers, Box 1, Correspondence, SCRC.
26 Henry Dreyfuss to John Vassos (16 November 1965), Vassos papers, Box 1, Correspondence, SCRC.
27 Arthur J. Pulos to William Katavolos (28 October 1965), Vassos papers, Box 1, Correspondence, SCRC.
28 IDSA Board of Directors Meeting (8 September 1968), Vassos papers, Box 2, Board of Directors, SCRC.
29 Anonymous response to IDSA questionnaire, IDSA Membership survey, Folder 4, Vassos papers, Syracuse, SCRC.
30 IDSA Executive committee minutes (7 January 1970), IDSA Archive, Syracuse, SCRC.
31 James Holland, *Minerva at Fifty: The Jubilee History of the Society of Industrial Artists and Designers 1930 to 1980* (Westerham: Hurtwood, 1980), p. 48.
32 In 1976, this was replaced with the Diploma membership and no longer carried an affix. By the 1990s, it was renamed 'Graduate' and declined dramatically, without the supporting mechanism of the Direct Admission Scheme. See Leah Armstrong, 'Designing a Profession: The Structure, Organization and Identity of the Design Profession in Britain, 1930-2010' (PhD thesis, University of Brighton, UK, 2014).
33 Geoffrey Millerson, *The Qualifying Associations: A Study in Professionalization* (London: Routledge, 1964), p. 91.
34 David Sugarman, *A Brief History of the Law Society* (London: Law Society, 1995), p. 5.
35 Thatcher was made Honorary Fellow of the CSD in 1982, in recognition of her efforts.
36 The Textile Institute had received its Royal Charter in 1910.
37 Dick Negus, *The Designer* (1977), Box 96, CSDA.
38 George Freeman to Terence Conran (5 December 1980), Box 93, CSDA.
39 Eddie Pond, Notes on meeting of Forward Planning Committee (27 June 1978), Box 100, CSDA.
40 Stefan Zachary, quoted in Helen Buttery, 'To Join or not to Join?', *Designer's Journal* (November 1984), Box 100, CSDA.
41 James Pilditch, 'Towards the Next Century: A Strategy for the Chartered Society of Designers', pp. 16–17, Box 13, CSDA. See Guy Julier, *Economies of Design* (London: Sage, 2017), p. 12.
42 'Club Class', *Blueprint* (May 1987), Box 100, CSDA.
43 Buttery, 'To Join or not to Join?'
44 June Fraser, interview with Robert Wetmore (15 May 1985), CSDA.
45 The grievance related to a *New York Times* article in which Stevens was credited for all Studebaker car products, including the Avanti, which Raymond Loewy had originally designed in 1953. Having brought the case to the SID grievance committee, Loewy stated that the committee 'dawdled' and Stevens refused to apologize, claiming that the fault was with the *New York Times*, Ethics/Grievance Folder, IDSA Archive, SCRC.

46 Victor Papanek to William Goldsmith, IDSA (1 August 1967), Correspondence, Box 6, Vassos papers, SCRC.
47 Alison Clarke, *Victor Papanek: Designer for the Real World* (Cambridge, MA: MIT Press, 2020), p. 278.
48 Victor Papanek, 'Preface', in *Design for the Real World: Human Ecology and Social Change* (New York: Pantheon, 1971), p. xxi.
49 Victor Papanek to William Goldsmith (23 June 1971), Vassos papers, SCRC.
50 According to Ray Spilman in a letter to the IDSA Board, 18 May 1971, 'several misunderstandings and misinterpretations occurred between Papanek and a Mr Taylor'. William Goldsmith finally replied to this letter in June 1971, having failed to notice it previously, remarking upon the 'misunderstandings', William Goldsmith to Victor Papanek (June 3 1971), Box 8 Correspondence, John Vassos Archive, SCRC.
51 Goldsmith claimed that the letter was lost by secretaries, probably to conceal the fact that they had not bothered to reply.
52 Raymond Spilman to Victor Papanek (19 May 1971), Box 8, Correspondence, Vassos papers, SCRC.
53 Ibid.
54 Victor Papanek to William Goldsmith (23 June 1971), Box 8, Vassos papers, SCRC.
55 See Chapter 1, p. 48–50.
56 Papanek quoted Vassos as saying, 'I just can't think of belonging to a society in which you are not a member with me' in his letter to Goldsmith, Victor Papanek to William Goldsmith (23 June 1971), Box 8, Vassos papers, SCRC.
57 Ibid.
58 William Goldsmith to Victor Papanek (14 July 1971), Box 8, Vassos papers, SCRC.
59 Ibid.
60 Clarke, *Victor Papanek*, p. 187. Papanek appears in the membership records for the IDSA (1965), IDSA Archive, SCRC.
61 Kaufmann Industrial Design Award Records (1910–1989), SCRC. This record states that the award was discontinued in 1967.
62 Victor Papanek to John Vassos (17 January 1964), Box 8, Vassos papers, SCRC.
63 Ibid.
64 Ibid.
65 See H. Alpay Er, Fatma Korkut and Ozlem Er, 'US Involvement in the Development of Design in the Periphery: The Case History of Industrial Design Education Turkey, 1950s–1970s', *Design Issues*, 19:2 (Spring 2003), 17–34.
66 Alison Clarke, 'Design for Development: ICSID and UNIDO: The Anthropological Turn in 1970s Design', *Journal of Design History*, 29:1 (2016), 43–57.
67 Maria Göransdotter, 'Designing Together: On Histories of Scandinavian User-Centred Design', in Kjetil Fallan, Christina Zetterlund and Anders V. Munch (eds), *Nordic Design Cultures in Transformation: Revolt and Resilience* (London: Routledge, 2023), pp. 157–77.
68 See Chapter 2 of this book on the values of teamwork, pp. XX.
69 In their history of the DRU, Avril and John Blake highlighted the significance of the Systematic Design Methods movement in the UK, which they claim was assimilated into the everyday practice of the DRU. As they state, 'one outcome of this attitude is the attempt to break down the barriers which more and more seem to divide society – scientist from artist, artist from designer, designer from engineer, engineer from the common man and so on in increasingly viscous circles'. Avril Blake and John Blake, *The Practical Idealists: Twenty-Five Years of Designing for Industry* (London: Lund Humphries, 1969), p. 55.
70 Jack Howe, 'The Responsibility of the Designer', Presidential Address at the RSA (25 September 1963), Jack Howe, RDI Box, FRDIA.
71 Bruce Archer, *Systematic Design Methods* (London: CoID, 1965).
72 Clarke, *Victor Papanek*, pp. 250–1.

73 Penny Sparke, *Consultant Design: The History and Practice of the Designer in Industry* (London: Pembridge Press, 1983), p. 83.
74 Greg Castillo, *Cold War on the Home Front: The Soft Power of Mid-Century Design* (Madison: University of Wisconsin Press, 2010).
75 See Jeremy Aynsley, Alison J. Clarke and Tania Messell (eds), *International Design Organizations: Histories, Legacies, Values* (London: Bloomsbury, 2022).
76 'Peripheral Vision: An interview with Gui Bonsiepe charting a lifetime of commitment to design empowerment' (2003), www.guibonsiepe.com/pdffiles/interview2_2003.pdf, p. 2, accessed 15 April 2024.
77 Daniel Magaziner, 'Designing Knowledge in Postcolonial Africa: A South African Abroad', *Kronos*, 41:1 (2015), www.scielo.org.za/pdf/kronos/v41n1/11.pdf, accessed 15 April 2024. I am grateful to Tania Messell for bringing this to my attention.
78 Undated letter/memo, inserted within the minutes for the IDSA meeting, 1972, with a letter attached from Professor R. Petrin, University of Nairobi. On 1 March 1966, Pulos wrote to Ramah Larisch at the IDSA to 'at long last clear up the question of Nathan Shapira's eligibility for membership of the IDSA', finding evidence that Shapira was a former member of the IDSA and as such, was entitled to IDSA membership. IDSA Archive, Folder 3, IDSA, SCRC.
79 Tucker Madwick to John Vassos (26 October 1972), Vassos papers, Box 1, Executive Committee Minutes, SCRC.
80 In August 1965, Arthur Pulos wrote to George Beck to express concern that the American Nathan Shapira would represent them at the Vienna ICSID conference, since he was 'not qualified and not a member of IDSA or IDEA'. Arthur J. Pulos to George Beck (20 August 1965), IDSA Archive, SCRC.
81 Tucker Madwick, President IDSA, Foreword, *Opportunities in Industrial Design Careers* (1970).
82 Arthur J. Pulos to George Beck (20 August 1965), IDSA Archive, SCRC, p. 16.
83 Ibid., p. 57.
84 Dorothy Goslett, 'Foreword', in *The Professional Practice of Design* (London: HarperCollins, 3rd edn, 1978).
85 Ibid.
86 Ibid., p. 157.
87 The eleven organizations were: the American Consulting Engineers Council, American Institute of Architects, American Institute of Graphic Arts, American Institute of Interior Designers, American Institute of Planners, American Society of Civil Engineers, American Society of Landscape Architects, Industrial Designers Society of America, National Society of Interior Designers, National Society of Professional Engineers and Package Designers Council. Bicentennial for American Design, Names and Addresses (March 1974). Folder 6, IDSA Archive.
88 George Nelson to William Goldsmith (March 1974), IDSA Archive, Folder 6, SCRC.
89 Ibid.
90 Ibid.
91 Handwritten letter to Roger Guilfoyle, Editor in Chief, *ID* magazine (n.d., c.1973), re: tribute to Henry Dreyfuss. Box 1, Correspondence, Folder 11, Vassos Papers, SCRC.
92 Gui Bonsiepe is referred to as a 'giant of the movement' (2003/2004 interview). Clarke, 'Design for the Real World', p. 98. Guy Julier refers to Bonsiepe as a 'hero' in 'Review Essay: Why Design Activists Need Historians', *Journal of Design History*, https://doi.org/10.1093/jdh/epac056, accessed 15 April 2024.
93 As Tania Messell explores, alternative models of social design were at play outside Western contexts – see Tania Messell, 'Globalization and Design Institutionalization: ICSID's XIth Congress and the Formation of ALADI, 1979', *Journal of Design History*, 32:1 (2019), 1–17.

94 Nathan Shapira Design Archive, material from 1968 to 2003. Topics covered include: sustainability, appropriate technology and localism, culture and identity politics of the image, universal and inclusive design, design for emerging economies, https://design.sfsu.edu/centers-archives, accessed 15 April 2024.
95 Clarke, *Victor Papanek*, p. 278. See also Alison J. Clarke, Amelie Klein and Matteo Kreis (eds), *Victor Papanek: The Politics of Design* (Weil am Rhein: Vitra Design Museum, 2018).
96 ICSID Conference: 'Design for Need, The Social Contribution of Design', RCA London, 1976. ICSID Archives, UBDA.
97 Lilián Sánchez Moreno, 'Towards Professional Recognition: Social Responsibility in Design Discourse and Practice from the Late 1960s to Mid-1970s' (PhD thesis, University of Brighton, UK, 2020).
98 Pauline Madge, 'Review Essay: Design for Society Re-makings: Ecology, Design, Philosophy, Nature, Technology, and Society: Cultural Roots of the Current Environmental Crisis, Interpreting Nature: Culture Constructions of the Environment', *Journal of Design History*, 7:4 (1994), 301–5.
99 Walter Schaer to Arthur J. Pulos (6 December 1975), IDSA Archives, SCRC.

Epilogue

> I feel that the profession has betrayed me. I do not feel particularly proud of how I make my living today. I think we were doing a good job. I'm quite comfortable with myself about what *we* are doing. I'm not comfortable with what I see my colleagues doing and I'm not comfortable with my colleagues. I think they are small people. I think they are little sort of opportunists and I do not see them thinking of the genuine significance of a human being and the things around him. I think we are dealing with a terribly important subject and we were not really giving a damn. I must say Ray, that, in retrospect ... my view of the field of design is that the, the airy-fairies that we used to scorn as being sort of doctrinaire, I think they are the ones that are holding it together today. I think the Charles Eames, George Nelson, Elliot Noyes – those people really are the last of the honourable designers ... I do not know many honourable designers today. They would rather trade a buck than have it right.[1]

Speaking in tone of resignation to his fellow designer Raymond Spilman, industrial designer Dave Chapman, past President of the ASID Chicago chapter, articulates the end of professionalization as a universal goal in industrial design. By the late 1970s, the mood of US industrial designers stood in stark contrast to the public buoyancy and optimism of the interwar period. It is fair to say that designers were not feeling very good about themselves or their contribution to society. In Britain too, commitment to the gentlemanly ideal of professionalism had faded dramatically. Membership of the CSD steadily declined and its relevance seemed increasingly outdated. In autumn 2001, design journalist and curator Stephen Bayley wrote in the *CSD Magazine*:

> Two generations ago a young Terence Conran had a little spat with the CSD because his self-promotional efforts were considered ungentlemanly and inappropriate to so stately a calling as a 'designer'. The subsequent and largely successful evolution of the design persona to its present pungent condition

of sundried narcissism shows that Conran was right and the old Society was wrong.[2]

Like Chapman, Bayley was identifying the limited hold of the professional ideal and the professional organization on the practice and identity of the industrial designer.

Design discourse, in practice, history and theory, continues to be propelled by an unending sense of crisis. Writing in 1993, industrial designer Adam Richardson declared the 'death of the designer', in which he called for a re-examination of the 'viability of the profession', its boundaries and values.[3] Design historians Craig Bremner and Paul Rodgers rearticulated this sense of crisis in 2013, in which they noted a 'crisis of professionalism, crisis of technology and crisis of economy' as some of the features that animate design in the twenty-first century. 'The design world today is a challenging and dynamic arena where professional boundaries are continually blurring', they state.[4] Indeed, it could be argued that designers, design historians and researchers have a particular propensity for 'crisis-making', engaging in a now longstanding tradition of apocalyptic statements about the status of the profession.[5]

Beyond the rhetoric, industry surveys, magazines and blogs in Britain and the US provide plenty of evidence on which to base this pessimism. Here, precarious work conditions, characterized by unpaid work, 'free-pitching', long working hours, demanding client expectations, marked gender imbalance and infringement of intellectual property are just some of the complaints raised.[6] Responding directly to these conditions, 'Precarity Pilot', 'an online platform and series of nomadic workshops that aim at addressing in inventive ways issues faced by precarious designers', focuses on the context of Europe, 'where cuts to welfare systems and unfair working conditions are making it difficult to confidently imagine the course of one's working life'.[7] At the Milan Triennale 2018, industrial design students from Eindhoven University used their work to critically question the role of the designer in precarious labour, automation and gentrification, in a series of works entitled '#NotForSale'.[8]

In the US, in 2013 the IDSA published the results of its research report, 'valuing the art of industrial design', in which it noted that fewer designers were salaried in industrial design than in other design fields, leaving them more vulnerable to exploitation.[9] Designers rarely blame the failure of design organizations to represent and enforce professional standards in design or embed any lasting legacy of ethical practice or minimum code of conduct. More often, they situate these problems in the context of volatile neoliberal market environments. While frustration may be rife, for some, it is precisely this irregularity that makes design such an attractive field of work. Some designers see value in the freedom of being freelance, and glamour and excitement in the unpredictability of the 24/7 creative lifestyle celebrated in the early representations of the Consultant Designer.[10] Online social media and video content-sharing platforms, including TikTok, Instagram and YouTube, overflow with self-created 'day in the life of a designer' content, in which individuals, a major proportion of whom are women, eulogise the 'designer lifestyle', and are often seen working from home, in ways that fascinatingly extend the narratives offered by mid-century Consultant Designers described in this book.

The claim, 'design is not social enough' has become a common refrain in contemporary practice, although it could be argued that designers remain critically disengaged from questions of what 'the social' actually consists of.[11] Designers are clearly awake to the crisis of professionalism, although, as design historian Kaisu Savola has recently noted, some of the urgency to find solutions, which characterized 1960s and 1970s Finnish 'social design' practices, or activism, at the student protests in Aspen, seems to have evaporated.[12] More positively, recent scholarship in 'decolonizing design' and 'depatriarchizing design' has punctured some of the 'heroic' and 'pioneering' narratives of the profession and shifted attention from the centre to the periphery – opening up new questions and new approaches to race, class and gender in highly productive ways.[13] Most fascinatingly, this work has been characterized by a dialogue between design historians and practitioners, highlighting a newly invigorated space between profession and research.

Broadly speaking, the design profession continues to hold a fractious and highly suspicious attitude to professionalism, professional identity and the values of professionalization. In 2010, the British design blog 'Creative Review' hosted an online discussion between designers on the value of professionalism in design, under the ominously familiar question, 'Is Design a Profession?' As one designer succinctly put it:

> Designers don't need to be professional ... Being certified just isn't very cool. This may sound frivolous but there are very many small design practices who will look on the idea of 'certification' with horror. Designers are not natural joiners and may prefer to try to raise the status of what they do through a less prescriptive, formal approach.[14]

This resistance to the very idea of being a professional is a theme that echoes through this book and in the pages of design history more broadly, since the formative years of the profession in both Britain and the US. Speaking in 1971, British design consultant FHK Henrion remarked, 'Young though the design profession is, in my own lifetime it has changed its function and description several times.'[15] Henry Dreyfuss, one of the founding members of the SID and a passionate proponent of professionalism in design, also acknowledged this fact:

> This tendency among designers to rejig, remodel, streamline and change as our ties change very likely is an occupational peculiarity. It's practically

compulsive and, although being industrial designers we may be a little bit biased, we think it's a positive peculiarity, a constitutive characteristic that gives us and our profession distinctive value.[16]

What Dreyfuss is coming to appreciate here is the unique value such a peculiarity can hold, since it means that the industry can adapt, move and respond to cultural and social change in a more fluid manner than if it was structurally encoded. As an industry founded in the intermediary space between producer and consumer, this was an essential characteristic. In many ways, the designer's failure to become fully recognised as a professional, its 'semi-professional status', was an asset as design culture moved into the next century.

Industrial design, like many other professions, seems to be stuck in an eternal struggle between its self-image and aspiration; how the profession sees itself and how it wishes to be seen. Starting from a position of anonymity and insecurity at the beginning of the 1930s, a cluster of design organizations, both governmental and voluntary, set out to invent a set of definitions, titles, categories and personas through which the identity of the design could be regularized, industrialized, professionalized and made visible. The establishment of the Society of Industrial Artists (SIA) in Britain and the Society of Industrial Designers in the US (SID) presents some evidence of the emergence of a professional consciousness in design, but neither was able to articulate or offer a stable definition of the designer's role that would last the century. The patronizing and patriarchal conviction that someone needed to 'tell young designers how to behave and what to do', as Dorothy Goslett recalls it, now appears to have been spectacularly misguided.

If Chapters 5 and 6 of this book documented this failure, the first four chapters diagnosed the conditions on which the profession was set up to fail. These included a superficial interpretation and application of the professional ideal, which was accepted in spirit but not always performed in practice; a pragmatic approach to professionalism that did not fit with the profession's commercial objectives and dependence on planned obsolescence, especially in the US; a moralistic drive characterized by the patronizing aims of the design reform movement in Britain. However, the professional designer also fell victim to wider industrial and social shifts beyond the control of institutions and organizations. This includes the shift from consultancy model to integration, the absorption of design work as part of a complete corporate identity service and the rise of 'creativity' as a competing cultural identity in design practice. Other factors included the fractious shifts in society broadly in the 1960s and 1970s, through the opening of competitive international markets, cold-war politics and environmental activism. Despite many efforts to impose a fixed set of ideals and prescribed codes in the profession, there has never commonly accepted ideal about what the role of the industrial designer should be in either place. As this book has shown, the industrial design profession has been necessarily inventive, continually designing and redesigning its role and place in the cultural economy. As a consequence, the designer's value, status and identity have never been fixed or secured, caught in the endless promise of a 'new profession'.

Notes

1. Dave Chapman, interview with Raymond Spilman (c.1964), Raymond Spilman Archive, interview tapes, SCRC.
2. Stephen Bayley, *CSD Magazine* (May 2001), 98, CSDA.
3. Adam Richardson, 'The Death of the Designer', *Design Issues*, 9:2 (Autumn 1993), 34–43.
4. Craig Bremner and Paul Rodgers, 'Design Without Discipline', *Design Issues*, 29:3 (Summer 2013), 4.
5. Leah Armstrong, Keynote Paper: 'Working in Crisis-Mode: Professionalism, Power and Precarity in Design', International Research Exchange Day, 'For the Sake of Creative Freedom', Hochschüle der Kunste, Bern, HKB (1 December 2023).
6. Design Industry Voices Report, 2012, www.designindustryvoices.com, accessed 5 April 2023.
7. Bianca Elzenbaumer, http://precaritypilot.net, accessed 5 April 2023. See also Bianca Elzenbaumer, 'Design(ers) Beyond Precarity: Proposals for Everyday Action', in Claudia Mareis and Nina Paim (eds), *Design Struggles, Intersecting Histories, Pedagogies and Perspectives* (Amsterdam: Valiz, 2021), pp. 317–30.
8. '#NotForSale', *Dezeen*, www.dezeen.com/2018/05/04/design-academy-eindhoven-students-design-interventions-milan-design-week (4 May 2018), accessed 15 April 2024.
9. Bonnie Nichols, 'Valuing the Art of Industrial Design: A Profile of the Sector and its Importance to Manufacturing, Technology and Innovation', IDSA 2013 Report (August 2013).
10. Jonathan Crary, *24/7: Late Capitalism and the Ends of Sleep* (New York: Verso, 2015); Angela McRobbie, *Be Creative* (London: Polity Press, 2015).
11. Craig Martin, *Deviant Design* (London: Bloomsbury, 2023).
12. Kasiu Savola, 'Disrespectful Thoughts about Design: Social, Political and Environmental Values in Finnish Design, 1960–1980' (PhD thesis, Aalto University, Finland, 2023).
13. Decolonising Design, www.decolonisingdesign.com, accessed 23 January 2024. Depatriarchise Design, https://depatriarchisedesign.com, accessed 23 January 2024.
14. '#NotForSale'.
15. FHK Henrion, speech at Maidstone, UK (1971) in FHK Henrion speeches, Folder 50, UBDA.
16. Henry Dreyfuss, IDSA *Design Notes*, 1:1, 3, Box 43, Arthur J. Pulos Archive, AAA.

Select bibliography

Archival sources

Archives of American Art, Smithsonian Institution, Washington, DC (AAA)
Cooper Hewitt, Smithsonian Design Museum Archives, New York (CHMA)
Cooper Hewitt, Smithsonian Design Museum National Design Library, Rare Books, New York (CHMRB)
Chartered Society of Designers Archive, London (CSDA)
Erno Goldfinger Archive, London, c/o National Trust (EGA)
Faculty of Royal Designers for Industry Archives, Royal Society of Arts, London (FRDIA)
Hirschl & Adler Galleries, New York, private archives (HAG)
Hagley Museum Archives, Wilmington, Delaware (HMA)
Jack Howe, personal archive, London, c/o Susan Wright (JHA)
National Museum of American History Archives Center, Smithsonian Institution, Washington, DC (NMAH)
Paul J. Getty Archives, Los Angeles, CA (PGA)
RIBA Archives, V&A Museum, London (RIBA)
Special Collections Research Center, Syracuse University Libraries (SCRC)
University of Brighton Design Archives, Brighton (UBDA)
Archives of Art and Design, Victoria & Albert Museum, London (VAAD)

Oral histories

Interviews by the author

Conran, Sir Terence (14 July 2011)
Garland, Ken (26 February 2013)
Pearson, David (21 March 2011, 3 February 2012)

Select bibliography

CSD Archive, London, Box 38: a collection of taped interviews by Robert Wetmore

Carrington, Noel (15 March 1984)
Carter, David (13 June 1984)
Collins, Jesse (27 March 1982)
Conran, Terence (24 June 1984)
Fraser, June (15 May 1985)
Goslett, Dorothy (23 January 1983)
Harris, David (14 April 1984)
Hatch, Peter (2 December 1981)
Henrion, FHK (13 Feb 1981)
Howe, Jack (17 March, 1981)
Julius, Leslie (28 June 1985)
Loewy, Raymond (n.d.)
Lord, Peter (5 April 1982)
Middleton, Michael (24 November 1984)
Reilly, Paul (31 May 1984)
Russell, Sir Gordon (29 July 1976)
Ward, Neville (19 March 1981)

Raymond Spilman taped interviews, Box 92, Raymond Spilman Archive, SCRC

Memorial, Henry Dreyfuss New York C w/Goldsmith, Hewitt, Jo Mezinger, Griffith, Nelson Formative years of USID (15 November 1973)
Chapman, Dave (28 December 1977)
Dreyfuss, Henry (8 December 1968)
Kibble, Ed/Edward, WDTA Memories (24 September 1990)
Laytham, Richard, Latham (1969)
Muller-Munk, Peter (n.d.)
Noyes, Eliot (n.d.)
Pulos, Arthur J., Pulos, 5 (c.1968)
Reinecke, Jean, Tape 36 (1979)
Spilman, Ray (4–5 February 1978)
Teague, Jr, Walter Dorwin (n.d.)
Vassos, John (13 July 1978)

Primary sources

Archer, Bruce (1965). *Systematic Method for Designers* (London: CoID).
The Architect and His Office: A survey of organization, staffing, quality of service and productivity presented to the council of the RIBA (6 February 1962). National Art Library, Victoria & Albert Museum.
Barr, Alfred H. (1934). *Machine Art* (New York: Museum of Modern Art).
Blake, Avril (1983). *The Black Papers on Design: Selected Writings of the Late Sir Misha Black* (Oxford: Pergamon Press on Behalf of the Faculty for Royal Designers for Industry).

Select bibliography

Blake, Avril (1986). *Milner Gray* (London: Design Council).
Blake, Avril and Blake, John (1969). *The Practical Idealists: Twenty-five Years of Designing for Industry* (London: Lund Humphries).
Brook, Tony and Shaughnessy, Adrian (2013). *Ken Garland: Structure and Substance* (London: Unit Editions).
Brook, Tony and Shaughnessy, Adrian (2014). *FHK Henrion: The Complete Designer* (London: Unit Editions).
Dreyfuss, Henry (1955). *Designing for People* (New York: Simon & Schuster).
Farr, Michael (1966). *Design Management* (London: Hodder & Stoughton).
Garland, Ken (1963). *First Things First: A Manifesto* (London: Ken Garland).
Gloag, John (1944). *The Missing Technician in Industrial Production* (London: Routledge).
Goslett, Dorothy (1960). *The Professional Practice of Design* (London: Batsford).
Greenough, Horatio (1957). *Form and Function: Remarks on Art, Design and Architecture* (Berkeley: University of California Press).
Harris, Jennifer (1993). *Lucienne Day: A Career in Design* (Manchester: Whitworth Art Gallery).
Henrion, FHK (1974). 'Design's Debt to Ashley', *Penrose Annual*, 67 (1974).
Holland, James (1980). *Minerva at Fifty: The Jubilee History of the Society of Industrial Artists and Designers 1930–1980* (London: Hurtwood Press).
Hunter, Alec B. (1957). 'The Anonymous Designer', *The Times* (14 January), p. 9.
Lippincott, Gordon (1947). *Design for Business* (Chicago: Paul Theobald).
Loewy, Raymond (1953). *Never Leave Well Enough Alone* (New York: Simon & Schuster).
Lonsdale-Hands, Richard (1957). Letter, 'The Anonymous Designer', *The Times* (16 January), p. 9.
McMahon, Sir Henry (1936). 'Institution of a Distinction for Designers for Industry', Dinner at the Society's House, 13 November, *RSA Journal* (November), pp. 2–10.
Mercer, F. A., Editor of Art and Industry (1945), Sixteenth Ordinary Meeting, Wednesday 11 April 1945, 'The Industrial Design Consultant', *Journal of the Royal Society of Arts* (8 June).
Minale, Marcello (1991). *How to Win a Successful Multi-Disciplinary Design Company* (London: Elfande).
Mumford, Lewis (1963 [1934]). *Technics and Civilization* (New York: Harcourt, Brace).
Nelson, George (1934). 'Both Fish and Fowl', *Fortune* (February), pp. 48–90.
Nelson, George (1960). *Problems of Design* (New York: Whitney).
Obituary (1977). 'Eliot Noyes, Industrial Designer of Many Familiar Products, Dies', *New York Times* (19 July).
Papanek, V. (1971). *Design for the Real World: Human Ecology and Social Change* (New York: Pantheon).
Pearson, David (1969). 'Are Designers Professional?', *Design Dialogue* (Spring), 15–16.
Special Correspondent (1957). 'The Anonymous Designer, Seeking a New Status in Industry', *The Times* (9 January).
Teague, Walter Dorwin (1940). *Design This Day, The Technique of Order in the Machine Age* (New York: Harcourt, Brace, 1940).
Tomrley, C. G. (1969). *Let's Look at Design* (London: Frederick Muller).

Van Doren, Harold (1940). *Industrial Design: A Practical Guide* (New York: McGraw Hill).
Wilkins, Bridget (1991). Interview with Milner Gray, *Typographica*, 40:1 (Winter), 6–8.
Woolman-Chase, Edna (1939). 'Come to the World's Fair and Go Back a Woman of Tomorrow', *Vogue* (February), n.p.

Secondary sources

Abbott, Tony (1998). *The System of Professions: An Essay on the Division of Expert Labour* (Chicago: University of Chicago Press).
Adamson, Glenn, Riello, Giorgio and Teasley, Sarah (eds) (2011). *Global Design History* (London; New York: Routledge).
Adams Stein, Jesse (2016). *Hot Metal, Material Culture and Tangible Labor* (Manchester: Manchester University Press).
Adams Stein, Jesse (2016). 'Masculinity and Material Culture in Technological Transitions from Letterpress to Offset Lithography, 1960s–1980s', *Technology and Culture*, 57:1, 24–53.
Alpay Er, H. (1997). 'Development Patterns of Industrial Design in the Third World: A Conceptual Model for Newly Industrialised Countries', *Journal of Design History*, 10:3, 293–307.
Araujo, Ana (2021). *No Compromise: The Work of Florence Knoll* (Princeton, NJ: Princeton University Press).
Armstrong, Leah (2016). 'Steering a Course between Professionalism and Commercialism: The Society of Industrial Artists and the Code of Conduct for the Professional Designer, 1945–1970', *Journal of Design History*, 29:2, 161–80.
Armstrong, Leah (2017). 'A New Image for a New Profession: Self-Image and Representation in the Professionalization of Design in Britain, 1945–1960', *Journal of Consumer Culture*, 19:1, 104–24.
Armstrong, Leah (2018). 'Fashioning the Designer in Britain: Working from Home, 1945–1960' in G. Julier et al. (eds), *Design Culture: Object, Practice, Profession* (London: Bloomsbury Academic).
Armstrong, Leah (2021). 'Fashions of the Future: Fashion, Gender and the Professionalization of Industrial Design, *Design Issues*, 37:3, 5–17.
Armstrong, Leah and McDowell, Felice (eds) (2018). *Fashioning Professionals: Identity and Representation at Work in the Creative Industries* (London: Bloomsbury Academic).
Atkinson, Harriet and Clarkson, Verity (eds) (2017). Special Issue: 'Design as an Object of Diplomacy Post 1945', *Design and Culture*, 9:2.
Atkinson, Paul (ed.) (2006). Special Issue: 'Do It Yourself: Democracy and Design', *Journal of Design History*, 19:1, 1–10.
Atkinson, Paul and Beegan, Gary (2006) 'Professionalism, Amateurism and the Boundaries of Design', *Journal of Design History*, 21:4, 305–13.
Attfield, Judith (2003). 'Review Essay: What Does History Have to do With It? Feminism and Design History', *Journal of Design History*, 16:1, 77–87.
Aynsley, Jeremy, Clarke, Alison J. and Messell, Tania (2021). *International Design Organizations: Histories, Legacies, Values* (London: Bloomsbury).
Barbieri, Chiara (2024). *Graphic Design in Italy: Culture and Practice in Milan, 1930–1960* (Manchester: Manchester University Press).

Bellini, Andrea and Maestripieri, Lara (2018). 'Professions Within, Between and Beyond: Varieties of Professionalism in a Globalising World', *Cambio*, 8:1, 5–14.
Bilbey, Diane (ed.) (2019). *Britain Can Make It: The 1946 Exhibition of Modern Design* (Paul Holberton Publishing/V&A Museum).
Blake, John and Blake, Avril, *DRU* (1964). *The Practical Idealists* (London: Lund Humphries).
Blakinger, John R., *Gyorgy Kepes, Undreaming the Bauhaus* (Cambridge, MA: MIT Press).
Bourdieu, Pierre (1982) *Language and Symbolic Power* (Cambridge, MA: Harvard University Press).
Bourdieu, Pierre (1984 [1979]) *Distinction: A Social Critique of the Judgement of Taste*, trans. Richard Nice (Cambridge, MA: Harvard University Press).
Bourke, Joanna (1996), 'The Great Male Renunciation: Men's Dress Reform in Inter-War Britain', *Journal of Design History*, 9:1, 23–33.
Breakell, Sue and Whitworth, Lesley (2015). 'Émigré Designers in the University of Brighton Design Archives', *Journal of Design History*, 28:1, 83–97.
Bremner, Craig and Rodgers, Paul, Craig Bremner and Paul Rodgers (2013). 'Design Without Discipline', *Design Issues*, 29:3, 4.
Breward, Christopher (1999). *The Hidden Consumer: Masculinities, Fashion and City Life* (Manchester: Manchester University Press).
Buchanan, Richard (1990). 'Toward a New Order in the Decade of Design', *Design Issues*, 6:2, 70–80.
Buckley, Cheryl (1986). 'Made in Patriarchy: Toward a Feminine Analysis of Women and Design', *Design Issues*, 3:2, 3–14.
Buckley, Cheryl (1990). *Potters and Paintresses: Women Designers in the Pottery Industry, 1870–1955* (London: The Woman's Press).
Buckley, Cheryl (2019). 'Made in Patriarchy II: Researching (or Re-searching) Women and Design', *Design Issues*, 6:1, 19–29.
Calvera, Anna (2005). 'Local, Regional, National, Global and Feedback: Several Issues to be Faced with Constructing Regional Narratives', *Journal of Design History*, 18:4, 371–83.
Campbell, Colin (1987). *The Romantic Ethic and the Spirit of Modern Consumerism* (Oxford: Blackwell).
Carrington, Noel (1976). *Industrial Design in Britain* (London: Allen & Unwin).
Carr-Saunders, A. M. and Wilson, P. A. (1933) *The Professions* (Oxford: Clarendon Press).
Castillo, Greg (2010). *Cold War on the Home Front: The Soft Power of Mid-Century Design* (Duluth: University of Minnesota Press).
Clarke, Alison J. (2014). '"Actions Speak Louder": Victor Papanek and the Legacy of Design Activism', *Design and Culture*, 5:2, 151–68.
Clarke, Alison J. (2016). 'Design for Development: ICSID and UNIDO: The Anthropological Turn in 1970s Design', *Journal of Design History*, 29:1, 43–57.
Clarke, Alison J. (ed.) (2017 [2012]). *Design Anthropology: Object Cultures in Transition* (London: Bloomsbury).
Clarke, Alison J. (2020). *Victor Papanek: Designer for the Real World* (Cambridge, MA: MIT Press).
Clarke, Alison J., Klein, Amelie and Kreis, Matteo (eds) (2018). *Victor Papanek: The Politics of Design* (Weil am Rhein: Vitra Design Museum).

Clarke, Alison J. and Shapira, Elana (eds) (2017). *Émigré Cultures in Architecture and Design* (London: Bloomsbury).
Crary, Jonathan (2013). *24/7: Late Capitalism and the Ends of Sleep* (London: Verso).
Darling, Elizabeth and Whitworth, Lesley (2017). *Women and the Making of Built Space in England* (Cambridge, MA: Belknap Press of Harvard University Press).
De Grazia, Victoria (2009). *Irresistible Empire: America's Advance Through Twentieth Century Europe* (Cambridge, MA: Belknap Press of Harvard University Press).
Durkheim, Émile (1992) [1957]. *Professional Ethics and Civic Morals*, trans. C. Brookfield (London: Routledge).
Durkheim, Émile (1997) [1893]). *Division of Labour in Society* (New York: Simon & Schuster).
Elfline, Ross (2016). 'Superstudio and the "Refusal to Work"', *Design and Culture*, 8:1, 55–77.
Evetts, Julia (2012). 'Similarities in Contexts and Theorising: Professionalism and Inequality', *Professionalism and Professions*, 2:2, 1–15.
Eyal, Gil (2019) *The Crisis of Expertise* (New York: Polity Press).
Fallan, Kjetil and Lees-Maffei, Grace (eds) (2016). *Designing Worlds: National Design Histories in an Age of Globalization* (Oxford: Berghahn).
Ferguson, Michael (2002). *The Rise of Management Consulting in Britain* (London: Routledge),
Florida, Richard (2002). *The Rise of the Creative Class* (New York: Basic Books).
Frayling, Christopher (1999). *Art and Design, 100 Years at the Royal College of Art* (London: Royal College of Art).
Göransdotter, Maria (2022). 'Designing Together, On Histories of Scandinavian User-Centered Design', in Kjetil Fallan et al., *Nordic Design Cultures in Transformation* (London: Routledge), pp. 157–177.
Gordon, Alastair (2003). *Andrew Geller: Beach Houses* (Princeton, NJ: Princeton University Press).
Gordon-Fogelson, Robert, 'Vertical and Visual Integration at Container Corporation of America', *Journal of Design History*, 35:1, 70–85.
Haber, Samuel (1991). *The Quest for Authority and Honor in the American Professions, 1750–1900* (Chicago: University of Chicago Press).
Hall, Catherine and Davidoff, Leonore (1987). *Family Fortunes: Men and Women of the English Middle Class 1780–1850* (London: Routledge).
Herring, Eleanor (2016). *Street Furniture Design, Contesting Modernism in Post-War Design* (London: Bloomsbury).
Hesmondhalgh, David (2002). *The Cultural Industries* (London: Sage).
Hobsbawm, Eric and Ranger, Terence (1983). *The Invention of Tradition* (Cambridge: Cambridge University Press).
Hudson, Liam (1992). *The Way Men Think: Intellect, Intimacy and the Erotic Imagination* (New Haven, CT: Yale University Press).
Hughes, Everett Cherrington (1963). 'Professions', *Daedalus*, 92:4, 655–68.
Illich, I. (1977). *Disabling Professions* (London: Marion Boyars).
Jackson, Peter A., Lowe, Michelle, Miller, Daniel and Mort, Frank (eds) (2000). *Commercial Cultures, Economies, Practices, Spaces* (London: Berg).
Jacobs, Meg (1999). 'Democracy's Third Estate, New Deal Politics and the Construction of a "Consuming Public"', *International Labour and Working Class History*, 55.

Jacobs, Meg (2005). *Pocketbook Politics, Economic Citizenship in Twentieth Century America* (Princeton, NJ: Princeton University Press).

Jerlei, Triin (2020). 'Developing Design Discourse in Soviet Print Media: A Case Study Comparing Estonia and Lithuania, 1959–1968', *Design and Culture*, 12:2, 203–26.

Julier, Guy (2015). *The Culture of Design* (London: Sage).

Julier, Guy (2017). *Economies of Design* (London: Sage).

Kauffman-Buhler, Jennifer (2022). 'Review: "No Compromise: The Work of Florence Knoll"', *Journal of Design History*, 35:2, 201–2.

Kaye, Barrington (1960). *The Development of the Architectural Profession in Britain* (London: Allen & Unwin).

Kaygan, Pinar (2017). 'Gender, Technology and the Designer's Work: A Feminist Review', *Design and Culture*, 8:2, 235–52.

Kelly, Jessica (2018). 'Introduction', Special Issue: 'Behind the Scenes: Anonymity and Hidden Mechanisms in Design and Architecture', *Architecture and Culture*, 6:8 (2018).

Kelly, Jessica (2022). *No More Giants: J. M. Richards, Modernism and the Architectural Review* (Manchester: Manchester University Press).

Kinross, Robin (1988). 'Herbert Read's Art and Industry: A History', *Journal of Design History*, 1:1, 35–50.

Kinross, Robin (1990). 'Émigré Graphic Designers in Britain Around the Second World War and Afterwards', *Journal of Design History*, 3:1, 35–57.

Kirkham, Pat (1996). *The Gendered Object* (Manchester: Manchester University Press).

Komlosy, Andrea (2018). *Work: The Last 1,000 Years* (London: Verso Books).

Laucht, Christopher (2020). Introduction, 'Transnational Professional Relations in the Long 19th Century', *Cultural and Social History*, 17:1 1–2.

Lees-Maffei, Grace (2008). 'Introduction: Professionalization as a Focus in Interior Design History', *Journal of Design History*, 21:1 1–18.

LeMahieu, D. L. (1988). *A Culture for Democracy, Mass Communication and the Cultivated Mind in Britain Between the Wars* (Oxford: Clarendon Press).

Lesko, Jim (1997). 'Industrial Design at Carnegie Institute of Technology, 1934–1967', *Journal of Design History*, 10:3, 269–92.

Lie, Ida Kamilla (2016). 'Make Us More Useful to Society!' The Scandinavian Design Students' Organization (SDO) and Socially Responsible Design, 1967–1973, *Design and Culture*, 8:3, 327–61.

Logemann, Jan (2017). *Engineered to Sell, European Emigres and the Making of Consumer Capitalism* (Chicago: University of Chicago Press).

MacCarthy, Fiona (1972) *All Things Bright and Beautiful: Design in Britain, 1830 to Today* (Oxford: Allen & Unwin).

MacCarthy, Fiona (1986) *Eye for Industry* (London: Royal Society of Arts).

Madoff, Steven H. (2009). *Art School, Propositions for the 21st Century* (Cambridge, MA: MIT Press).

Maestripieri, Lara (2016). 'Professionalization at Work: The Case of Italian Management Consultants', *Ephemera: Theory and Politics in Organization*, 16:2, 31–52.

Maffei, Nicholas P. (2018). *Norman Bel Geddes* (London: Bloomsbury).

Maguire, Patrick and Jonathan Woodham (1997) (eds), *Design and Cultural Politics in Postwar Britain: The Britain Can Make It Exhibition of 1946* (Leicester: Leicester University Press).

Manzini, Ezio (2015). *Design: When Everybody Designs* (Cambridge, MA: MIT Press).

Marchand, Roland (1985). *Advertising the American Dream: Making Way for Modernity, 1920–1940* (Berkeley: University of California Press).

Marchand, Roland (1993). 'The Designers go the Fair: Norman Bel Geddes, the General Motors "Futurama" and the Visit to the Factory Transformed', *Design Issues*, 8:2, 4–17.

Marchand, Roland (2001). *Creating the Corporate Soul, The Rise of Public Relations and Corporate Imagery in American Big Business* (California: University of California Press).

Mareis, Claudia and Paim, Nina (eds) (2021). *Design Struggles, Intersecting Pedagogies, Histories and Perspectives* (Amsterdam: Valiz).

Mazur Thomson, Eleanor (1997). *The Origins of Graphic Design in America, 1870–1920* (New Haven, CT: Yale University Press).

McBrinn, Joseph (2021). *Queering the Subversive Stitch, Men and the Culture of Needlework* (London: Bloomsbury).

McCarthy, Helen (2007). 'Parties, Voluntary Associations and Democratic Politics in Inter-War Britain', *Historical Journal*, 50:4, 891–912.

McKendrick, Neil and Outhwaite, R. B. (1986). *Business Life and Public Policy, Essays in Honour of D C Coleman* (Cambridge: Cambridge University Press).

McKenna, Christopher D. (2001). 'The World's Newest Profession: Management Consulting in the Twentieth Century', *Enterprise & Society*, 2:4, 673–9.

McQuiston, Liz (1988). *Women in Design: A Contemporary View* (New York: Rizzoli International).

Meikle, Jeffrey (2005). *Design in the USA* (Oxford: Oxford University Press).

Millerson, Geoffrey (1964). *The Qualifying Associations: A Study in Professionalization* (London: Routledge).

Moriarty, Catherine (2000). 'A Backroom Service? The Photographic Library of the Council of Industrial Design, 1945–1965', *Journal of Design History*, 13:1, 39–57.

Nenadic, Stena (2023). *Craftworkers in Nineteenth Century Scotland, Making and Adapting in an Industrial Age* (Edinburgh: Edinburgh University Press).

Nixon, Sean (1996). *Hard Looks: Masculinities, Spectatorship and Contemporary Consumption* (Oxford: Sage)

Nixon, Sean (2003). *Advertising Cultures, Gender, Commerce, Creativity* (Oxford: Sage).

Nixon, Sean (2014). *Hard Sell, Advertising, Affluence and Transatlantic Relations, 1951–1969* (Manchester: Manchester University Press).

Parsons, Talcott (1939). *The Professions and Social Structure* (Oxford: Oxford Academic Press).

Perkin, Harold (1989). *The Rise of Professional Society* (Chicago: University of Chicago Press).

Pilditch, James (1973). *The Silent Salesman: How to Develop Packaging that Sells* (London: Business Books).

Pilditch, James (1990). *'I'll be over in the morning': A Practical guide to Winning Business in Other Countries* (London: Mercury Books).

Pilditch, James and Scott, Douglas (1965). *The Business of Product Design* (London: Business Publications).

Preston, David (2011).'The Corporate Trailblazers', *Ultrabold*, Journal of St Bride Library, Summer.

Prokop, Ursula (2019). *Jacques and Jacqueline Groag: Two Hidden Figures of the Viennese Modernism Movement* (New York: DoppelHouse Press; illustrated edition).
Pulos, Arthur J. (1965). *American Design Ethic: A History of Industrial Design* (Cambridge, MA: MIT Press).
Pulos, Arthur J. (1970). *Opportunities in Industrial Design* (New York: Vocational Guidance Manuals).
Pulos, Arthur J. (1983). *The American Design Adventure* (Cambridge, MA: MIT Press).
Pursell, Carroll (1993). 'Am I a Lady or an Engineer?' The Origins of the Women's Engineering Society in Britain, 1918–1940', *Technology and Culture*, 34:I, 78–97.
Rappaport, Erica (1999). *Shopping for Pleasure: Women in the Making of London's West End* (New Haven, CT: Princeton University Press, 1999).
Roper, Michael (1994). *Masculinity and the British Organization Man since 1994* (Oxford: Oxford Academic Press).
Ross, Phyliss (2009). *Gilbert Rohde: Modern Design for Modern Living* (New Haven, CT: Yale University Press).
Rothschild, Joan (1999). *Design and Feminism: Re-visioning Spaces, Places and Everyday Things'* (New Brunswick, NJ: Rutgers University Press).
Saint, Andrew (1983). *The Image of the Architect* (New Haven, CT: Yale University Press).
Seddon, Jill (2000). 'Mentioned but Denied Significance: Women Designers and the Professionalization of Design in Britain c.1920–1951', *Gender and History*, 12:2, 426–7.
Seddon, Jill and Worden, Suzette (eds) (1994).*Women Designing: Redefining Design in Britain Between the Wars* (London: The Women's Press).
Serulus, Katarina (2018). *Design and Politics: The Public Promotion of Post-War Industrial Design in Belgium, 1950–1986* (Leuven: University of Leuven Press).
Sharma, Sarah and Sing, Rianka (2022). *Re-Understanding Media, Feminist Extensions of Marshall McLuhan* (Durham, NC: Duke University Press).
Sparke, Penny (1983). *Consultant Design: The History and Practice of the Designer in Industry* (London: Pembridge Press).
Sugarman, David (1995). *A Brief History of the Law Society* (London: Law Society).
Theofilou, Anastasios (ed.) (2021). *Women in PR History* (New York: Routledge).
Thomas, Zoë (2015). 'At Home with the Women's Guild of Arts: Gender and Professional Identity in London Studios: c.1880–1925', *Women's History Review*, 246: 938–64.
Thomas, Zoë and Egginton, Heidi (2002). *Precarious Professionals: Gender, Identities and Social Change in Modern Britain* (Chicago: University of Chicago Press).
Thompson, Christopher (2011). 'Modernizing for Trade: Institutionalising Design Promotion in New Zealand, 1958–1967', *Journal of Design History*, 24:3: 223–39.
Tickner, Lisa (2008). *Hornsey: 1968, The Art School Revolution* (London: Frances Lincoln).
Trachtenberg, Alan (1982). *Incorporation of America* (New York: Hill & Wang).
Twemlow, Alice (2009). 'I can't talk to you if you say that: An ideological collision at the International Design Conference at Aspen, 1970', *Design and Culture*, 1:1, p. 23–49.

Twemlow, Alice (2018). *Sifting the Trash, A History of Design Criticism* (Cambridge, MA: MIT Press).
Walker, Lynne (1984). *Women Architects: Their Work* (London: Sorella Press).
Wang, David and O'Ilhan, Ali (2009). 'Holding Creativity Together: A Sociological Theory of the Design Professions', *Design Issues*, 25:1, 5–21.
Wilson, Kristina (2021), *Midcentury Modernism and the American Body: Race, Gender and the Politics of Power in Design* (Princeton, NJ: Princeton University Press).
Woodham, Jonathan M. (1983). *The Industrial Designer and the Public* (Oxford: Pembridge Press).
Woodham, Jonathan M. (1997). *Twentieth Century Design* (Oxford: Oxford University Press).
Woodham, Jonathan M. (2010). 'Formulating National Design Policies in the United States: Recycling "The Emperor's New Clothes"?', *Design Issues*, 26:2, 27–46.
Woodham, Jonathan M. and Thomson, Michael (2017). 'Cultural Diplomacy and Design in the Late Twentieth and Twenty-First Centuries: Rhetoric or Reality?', *Design and Culture*, 9:2, 225–41.

Unpublished secondary sources

Armstrong, Leah (2014). 'Designing a Profession: The Structure, Organization and Identity of the Design Profession in Britain, 1930–2010' (AHRC Collaborative Doctoral Award, University of Brighton, UK).
Gordon-Fogelson, Robert (2021). 'Total Integration: Design, Business and Society in the United States, 1935–1985' (PhD thesis, University of Southern California).
Messell, Tania (2014). 'Constructing a United Nations of Design: ICSID and the Professionalization of Design on the World Stage, 1957–1980' (PhD thesis, University of Brighton, UK).
Preston, David (2019). 'The Logic of Corporate Communication Design' (PhD thesis, Central St Martins, University of the Arts London).
Sánchez Moreno, Lilián (2020). 'Towards Professional Recognition: Social Responsibility in Design Discourse and Practice from the late 1960s to Mid-1970s' (DPhil thesis, University of Brighton, UK).
Savola, Kaisu (2023). 'Disrespectful Thoughts about Design: Social, Political and Environmental Values in Finnish Design, 1960–1980' (PhD thesis, Aalto University, Finland).

Online and film sources

Boyd Davis, S., and Gristwood, S. (2016) 'The Structure of Design Processes: Ideal and Reality in Bruce Archer's 1968 Doctoral Thesis', in P. Lloyd and E. Bohemia (eds), *Proceedings of DRS2016 International Conference, Vol. 2: Future–Focused Thinking*, 27–30 June, Brighton, UK, Design Research Society, https://doi.org/10.21606/drs.2016, accessed 15 April 2024.
Bullmore, J. (2023), 'David Ogilvy: Life lessons from the godfather of advertising', *Gentleman's Journal* (14 April), www.thegentlemansjournal.com/article/david-ogilvy-life-lessons-godfather-advertising, accessed 4 May 2023.
Creative Review Blog (2012), www.creativereview.co.uk/cr-blog/2010/february/cr-survey (2 February), accessed 15 April 2014.

Design Council Report (2018) 'Designing a Future Economy' (London: Design Council) www.designcouncil.org.uk/fileadmin/uploads/dc/Documents/Designing_a_future_economy18.pdf, accessed 15 April 2024.

Design Industry Voices Report (2013) www.designindustryvoices.com/2013/infographics/design_industry_voices_report_2013, accessed 30 January 2015.

Dezeen, 'Design Academy Eindhoven Students Take Over Milanese Streets to Explore how Design Impacts Everyday Life' (4 May 2018), www.dezeen.com/2018/05/04/design-academy-eindhoven-students-design-interventions-milan-design-week, accessed 15 April 2024.

Julier, Guy, 'Disobedience as Usual: Why Design Activists Need Historians', *Journal of Design History*, https://doi.org/10.1093/jdh/epac056 (29 May 2023), accessed 15 April 2024.

Lonsdale-Hands, Richard (1961). Promotional video for the Lonsdale-Hands Organisation, Hirschl & Adler Galleries, New York, https://vimeo.com/28355748, accessed 15 April 2024.

Matter of *Teague* v. *Graves*, 261 Appellate Div. 652 (N.Y. App. Div. 1941), https://casetext.com/case/matter-of-teague-v-graves, accessed 15 April 2020.

'Peripheral Vision: An Interview with Gui Bonsepi: Charting a Lifetime of Commitment to Design Empowerment', www.guibonsiepe.com/pdffiles/interview2_2003.pdf, accessed 15 April 2024.

Precarity Pilot, http://precaritypilot.net, accessed 15 April 2024.

San Francisco State University Centres and Archives, Nathan Shapira Design Archive, https://design.sfsu.edu/centers-archives, accessed 15 April 2024.

Tonkinwise, Cameron (2015), 'Design for Transitions – From and to What?' Rhode Island School of Design, https://digitalcommons.risd.edu/cgi/viewcontent.cgi?article=1004&context=critical_futures_symposium_articles, accessed 15 April 2024.

Twemlow, A. (2007) 'The Consequences of Capitalism: An Interview with Ralph Caplan', https://alicetwemlow.com/the-consequences-of-criticism-an-interview-with-ralph-caplan, accessed 15 April 2024.

Index

administration 96, 111–13, 116, 118–20
Advertiser's Weekly 76
advertising
  'acceptable advertising' 140
  advertisements 73, 80, 106, 144, 181, 192
  see also profession: 'new profession'
Advertising Age 86
Allen, Deborah 95, 133
Allen, Margaret 77
American Artist 86
American Association of Advertising Agencies 45
'American Business and Industrial Design' report 72
'American design adventure' 43
American Designers Institute (ADI) 9, 47–9, 51–4, 57, 105, 118–19, 137
American Institute of Electrical Engineers 43
Americanization 11
anonymity
  'The Anonymous Designer' 80
anthropology
  'anthropological turn' 188
Architects Journal 98
Architectural Association (AA) 34, 135
Architectural Forum 111, 144
archives 16–17, 55, 69, 88, 96, 111, 115–16, 140

Arens, Egmont 50, 69, 74
Art and Industry 30, 64, 105–6, 118, 141
art director, 55, 135, 169, 170
Arts and Crafts movement 31
Aspen 118, 149, 153–8, 202
aspiration 2–3, 5–9, 14, 80, 97–8, 101–3, 129, 203
Atomic Energy Commission 179
Attfield, Judith 13
Austrian émigrés 78
Avanti 167
architecture, the profession 5, 6, 9, 15, 29, 34, 43, 80, 109, 133, 135, 138, 139, 144, 149, 158
The Architect and His Office 158

'backroom' administration 111
'Boy Scout' attitude 128, 139, 149
'Building a World of Tomorrow' *see* New York World's Fair
Bach, Alfons 47
Banham, Reyner 95, 155, 157
Bass, Saul 154, 171, 172
Bassett Gray Group of Artists and Writers 31
BecVar, Arthur 48–9, 71, 141
Belcher, W. L. 117
Bensusan, Geoffrey 164
beards 46, 132, 133, 149, 156
  see also moustaches

Index 217

Beresford-Evans, J. 112
Better Homes and Gardens 105
Bicentennial Project 192
Big Character Poster No1: Work Chart for Designers (1969) 187, 188
Black, Misha 9, 14, 34, 35, 37–42, 55–6, 76, 85, 88, 103, 113, 127, 132, 140, 146, 147, 164, 194
Bohemian 127
Bonney Leicester, Louise 106, 143
Boswell, James 129, 131, 132
Bourdeau, James 53
Brewer, Cecil 31
'Britain Can Make It' 39, 40, 41, 46, 54, 103, 171
British Institute for Industrial Art (BIIA), 30
Buchanan, Richard 83
Business Week 56, 66
The Business of Product Design 87

Caplan, Ralph 106, 109
Carrington, Noel 31, 42
Casson, Hugh 129
Ceramics 4, 13, 29, 30, 32–3, 53, 77, 97, 119
Chadwick, Hulme 76
Chapman, Dave 105, 138–9, 167, 178, 200, 201
Chartered Society of Designers (CSD) 16, 113, 115, 118, 162, 183, 192, 200
Chicago 43, 47, 48, 105, 200
Christmas cards 128, 140, 164
Clarke, Alison J. 178, 184, 186, 189, 193
class 5, 6, 10, 17, 29, 34, 40, 42, 79, 120, 129, 187, 202
climate of discontent 167–8
clients
  'client attention' 106, 141
  'designer-client relationship' 132, 136, 139, 177
  'how to be a good client' 160
  'The Industrial Designer and His Client', SID 138
  satisfied clients 141
  'Working with your designer', SIA 136, 139, 160

co-ordination
  designer as co-ordinator 188–90
Coates, Wells 36, 129
codes 127, 128, 135, 136
  British Code of Advertising Practice 164
  formal and informal 128
  code of conduct 128, 134–49, 158, 161, 162, 171, 190, 192, 201
  code of ethics 3, 10, 52, 137–9, 143, 144, 160, 166
  code of obligation 138–40
  gentleman's code 128
  international codes 145
cold war, the 5, 148, 157, 173, 178, 179, 189, 203
colonialism 179
commercialism 9, 15, 75, 85, 89, 163, 169, 172
company designer 70
Conran, Terence 12, 14, 88, 141, 142, 151, 164, 173, 182–3, 201
  Conran Design Group 88
consultancy
  Consultant Design (1983) 11
  consultant designer 17, 47, 49, 54, 57
  General Consultant Designer 64, 65, 77, 79, 80, 84, 85, 89, 90, 132, 145, 163
  'Big Four', the 105
consumer capitalism 45, 73, 104, 154
consumer culture 7, 14, 29, 37, 45, 74, 103
'Consumers in Danger' 55
Container Corporation of America (CCA) 81, *82*, 169
corporation
  'captive designers' 73
  corporate designer 48, 72, 167, 168, 172, 181
  corporate identity 104, 203
  corporate soul 11
  The Corporation and the Designer 155
Council for Art and Industry (CAI) 30, 118, 34

Index

Council of Industrial Design (CoID) 30, 37–8, 41–2, 55–6, 77, 98, 103, 115, 116, 127, 135, 140, 157
Coventry 33
Crawford, William 76, 117, 129
creative work 17, 94, 10 see also profession: 'creative profession'
creativity 25, 71, 118, 168–9, 170, 203
Cressonnieres de, Josine 120, 193
crisis 15
 of ecology 3
 environmental crisis 2
 of masculinity 132, 138, 149,
 of professionalism 153, 154, 157, 158, 165, 168, 172, 182, 201, 202
 see also 'climate of discontent'
Cuccio, John 166
cultural diplomacy 146, 189
cultural economy 169, 204

Dailey, Dave 139
Darling, Elizabeth 13, 41, 96
Darrach, Betsy 106
Darwin, Robin 117, 127
Daves, Jessica 138
Davies, Anne 85
Day, Lucienne 80, 120
Day, Robin 80, 157
de Holden Stone, James 36, 76
Deighton, Len 132, 133
De Mayo, Willy 76
De Pree, Hugh 178
Design and Art Directors Club (D&AD) 170, 183
Design and Industries Association (DIA) 9, 30, 31, 33, 47, 78
Design Council 94, 142, 183
Design Dialogue 145
Design disciplines 4
 Design for Craft-Based Industries 36
 Display, Furniture and Interior Design 36
 Design Direction 36
 Graphic Design 29, 54, 78, 129, 169, 170, 171
 Illustration 31, 34, 36, 46, 132
 product design engineering 78
 staff designer 70, 72, 109

Systematic Design Methods movement 93, 115, 189, 189
Television Film and Theatre 36
Textile and Dress Design 36
see also industrial design
'Design for Quality in Life' conference (1973) 189
Design for development 187, 189, 190, 191
Design for Need: the Social Contribution of Design, conference, (1976) 194
Design for the Real World (1960) 184, 189
Design magazine 48, 55, 56, 83, 84, 85, 114, 163
design management 84, 115
design manager 39
design methods movement 93, 115, 189, 189
Design Museum, the 183
design reform movement 30, 53, 203
Design Research Unit (DRU) 14, 39, 42, 76, 77, 87, 90, 112–15, 189, 192
Design Society of Kenya 190
design team 84, 187
 'The Minimal Design Team' 187, 188, 189
Design Week 114
Design: A Mid-Century Survey 55
designers
 'big-name' designers 69, 76, 90, 179
 'dangerous designer' 190
 'designer-craftsman' 77
 'designer lifestyle' 80, 202
 'Designers at Work' exhibition, (1957) 116
 'Designing Men' 43, 45
 elder designers 172
 'honourable designers' 200
 'The Designer and His Car' 129
 'The Designer', special issue, *SIA Journal* 129
 the systematic designer 83
 'young designers' 52, 114, 127, 163, 192, 203
Designers Journal 183
Deskey, Donald 46, 54, 73, 88, 191

Detroit 47, 48, 166
Deutsche Werkbund 9, 31
developing countries
　Working Committee Developing Countries 190
Diamond, Freda 14, 47, 96, 102–4, 120
Dickens, Ronald 161
Direct Admission Scheme (SIAD) 162–5, 170–3, 178, 181, 182
Doblin, Jay 144
Dohner, Donald 67, 68
Dorwin Teague, Walter 9, 12, 14, 48–9, 51, 67, 70–1, 74, 88, 90, 105, 136, 141, 143, 165, 184, 187
Double Crown Club 9, 31, 47, 140
Dralle, Elizabeth 119
dress 1, 36, 46, 117–18, 127, 132, 149
Dress Committee of the Council for Art and Industry 118
Drexler, Arthur 134
Dreyfuss, Henry 9, 12, 14, 45, 48–51, 67, 74, 83, 88, 103, 133, 134, 141, 144, 156, 165–7, 180–1, 187, 193, 202, 203
Du Pont 73
Durkheim, Émile 3, 5, 7, 8, 28, 36
Dutton, Norbert 38

Eames, Charles 49, 116, 173, 200
Eames, Ray 120
East Africa 190–1
Eastman Kodak 72, 103
Everett, Julian 180
egg cup 38, 40–2
émigré designers 11, 27, 34, 53, 78, 146
esthetics 181
ethics, professional 3, 9, 10, 48, 52, 74, 114, 136, 137, 143, 144, 160, 166, 181, 189
European Common Market 85–7, 147

'Form and Function' 46
'Forward Planning Committee' 183
Farr, Michael 55, 84, 115
Fashion 46, 51, 102, 104, 119
　'Fashions of the Future', *Vogue* (1939) 45, 138
Federal Design program 178

fee scales 183
Fejer, George 163
femininity 46, 96–100, 104, 120
　'feminine expertise' 102, 103, 106
First things first (1963) 170, 171
First World War 30
Fishbein, William 100
FORM 133
Fortune 43, 46, 51, 67, 74, 80, 82 87, 105
Franke, Ann 118
Fraser, Eric 32, 34
Fraser, June 97, 183, 184
Freedom 17, 71, 74, 148, 169, 192, 201
Freeman, George 183
Fuller, J. B. 164, 193
future, the
　'man of the next century' 46
　'woman of tomorrow' 45

Gardner, James 37–40, 42, 98
Garland, Ken 15, 169, 170
General Electric 71, 98, 143
generalism vs specialism model 187
　generalization and specialization; generalist and specialist skills 31, 65, 66, 70–78, 81, 84, 191, 193
generations
　'first generation' 85, 170, 191
　generational conflict 153, 156
　generational divide 157, 170
　see also maturity; 'young designers'; 'elder designers'
gentleman-architect 6, 17
gentleman-designer 86, 127
gentleman's club 115
'Gentleman's Code' 128
gentlemanly 6, 10, 16–17, 131–4, 38, 55, 115, 129, 142, 146, 163, 182, 192, 200
Gerth, Ruth 118
Gloag, John 37, 74, 75, 8
globalization 2, 3, 65, 88
Goldsmith, William 173, 185, 186, 192
good design 4, 37, 42, 53, 104, 138, 182
'Good Design and Good Business' 37
Good Housekeeping 106
Goddard, Lily 161

Gordon-Fogelson, Robert 154
Gordon-Miller, Leon 106
Goslett, Dorothy 39, 112–15, 117, 121, 132, 136, 140, 156, 162, 163, 192, 203
government reports 8, 37, 118
graduates 50, 117, 133
graphic design *see* Design disciplines
Gray, Milner 9, 14, 31–6, 55, 76, 77, 113, 135, 146, 160, 182
Great Revolt, the 46
Greenly's 86
Greenough, Horatio 46
grievances 137, 190
Groag, Jacqueline 78, 120
Guild, Laurelle 67, 68

Haber, Samuel 6, 66
'handicap program' 180
Harper's Bazaar 98
Harris, David 115, 142
Harrods 113
Havinden, Ashley 129, 135, 140
Heal, Ambrose 31
Heath, Alec 161
Height, Frank 161
Henrion, FHK 55, 76, 78, 79, 84, 90, 101, 116, 146, 147, 170, 202
Herman Miller 178
heroes 6, 10
The Hidden Persuaders (1957) 155
Hille 156, 171
Hoch, Ernest 161
Holland, James 78, 79, 118
Hollywood 79, 98
Holmes, Kenneth 76
home 13, 14, 27, 55, 95, 100, 101, 102–6, 109, 117, 120, 129, 202
Home Furnishings Market Conference 104
House and Garden 100, 106, 111
Housewife 14, 40, 96, 102, 103, 104, 120, 155
Howe, Jack 56, 136, 142, 163, 164, 189
'How to use your industrial designer', SID 140
Hudnut, Dean 52, 53
Human-centered design 184

'Human Factors' 178, 179
Hunter, Alec 80, 81

I'll be over in the morning (1990) 87, 160
Ibiza, (ICSID Conference) 190
Illich, Ivan 1
Illinois Institute of Technology 53, 95
Illustration, see Design disciplines
industrial art 29, 30, 32, 54
industrial artist 4, 29, 30, 31, 34, 36, 57
Industrial Design 15, 55, 57, 70, 83, 95, 98, 106, 133, 154, 155, 186, 191, 193
industrial design
  career 191
  definition – ICSID 145
  education 5, 45, 48, 50, 52, 53, 71, 76, 81, 145, 167
Industrial Design: A Practical Guide 114
industrial designer, profession
  'industrialized designer', the 4, 7–12
  'What and Why is an Industrial Designer', SID (1946) 139
  'What Industrial Design Means' *see* 'Britain Can Make It'
Industrial Design Partnership (IDP) 55, 76
Industrial Design Standards in England report 47
Institute of Contemporary Art, London (ICA) 84, 174
Institute of Electrical Engineers 135
  see also American Institute of Electrical Engineers
Institute of International Education 186
integration 65, 81, 82, 83, 84, 144, 169, 203
interior design 36, 51, 103, 139
Interiors 106, 111, 133
intermediaries 10
International Design Conference in Aspen (IDCA) 118, 149, 154–5, 158, 159, 166
International Council of Societies of Industrial Design (ICSID) 5, 128, 145–7, 149, 153, 162, 164, 177, 185, 189, 190, 193, 194

Index

internationalization 89
inter-war 178

Jacobs, Meg 42
Jacques, Robin 132
Jaguar 80
Jergensen, George A. 55, 56
job classification 69
Jones, Barbara 129, 161
Julier, Guy 2
Julius, Leslie 156

Katavolos, William 48, 181
Kaufmann International Design Awards 186
Kaufmann Jr., Edgar 136, 165
Kepes, György 38
Kinross, Robin 54
Knoll Bassett, Florence 16, 96, 101, 102, 120
Komlosy, Andrea 96
Khrushchev, Nikita 187

Lamb, Lynton 129, 172
Landor, Walter 55, 73, 76, 105, 184
Larisch, Ramah 146
law, profession 5, 6, 9, 29, 50, 66, 71, 135
  see also profession: 'older professions', 'traditional professions'
  law courts 33
Lessons, K. 117
'lessons unlearned' 178
licensing 5, 48, 53, 178, 180, 181
Licentiate membership 182
Life magazine 27
Lippincott–Margulies 143
Lippincott, Gordon 7, 10, 44, 71, 143
'lipsticks to locomotives' 71
Loewy, Raymond 9, 12, 14, 27, 48–50, 54, 56, 67–9, 71, 74, 75, 86, 87, 89, 100, 105–11, 120, 136–8, 143, 166, 167, 169, 173, 178, 179, 181, 184, 187
  Raymond Loewy Associates (RLA) 107, 108, 109, 129, 169
Logeman, Jan 11, 45, 66, 149
Lomas, Miss 116, 117

London 30–3, 36, 37, 55, 56, 64, 75, 79, 80, 84–8, 117, 127, 140, 145, 160, 161, 170, 183, 194
Lonsdale-Hands Organization 87
Lonsdale-Hands, Richard 76, 81, 85–7, 101
Los Angeles 48, 56, 166, 168
Luder, Owen 158

MacCarthy, Fiona 12
machine art 30
Madwick, Tucker 191
magazines 3, 10, 14, 15, 57, 79, 80, 95, 97, 98, 101, 104–6, 111, 177
Maldonado, Tomás 153, 162, 193
Malone, Robert 186
management consulting 66, 71
Manchester 33
Manchester Polytechnic 189
manifesto *see First Things First*
manipulation 55, 73, 155
Marchand, Roland 11, 73, 104
marriage
  'Wives' Coordinator' 119
  'Bridal School' 102
  'career couple' 102
  'Presidential wives' 118
Martin Senour Company 169
masculinity
  hyper-masculinity 12, 28, 38, 40, 121
  'Men of Design' 32
  see also designers: 'Designing Men' 43
mass production 8, 9, 31, 36–7, 45, 184
'matchstick to a city' 71
maturity 153, 185
  see also generation: generational divide
McBrinn, Joseph 12
McCall's 106
McConnell, Phillip 73, 106, 118, 138
McHugh, Ursula 106
McQuiston, Liz 98
medal
  SIAD medal 183
mediation, 2, 14–18, 120
  see also intermediaries
  'The Man in the Middle' (1958) 4, 154

Mercer, Frank A. 64, 74–6, 89, 90
mid-century 11, 14, 45, 53, 70, 102, 202
middle class 6, 40, 42, 155, 187
Middleton, Michael 160, 161
Meikle, Jeffrey 83
Milan Triennale, the 201
Mills, C. Wright 4, 120, 154, 184
misogyny 114
The Missing Technician 74
Modern Architectural Research Society 36
modernist design 14, 46
Morgan, N. 117
moustaches, see beards
  'M. Designer' 37–40, 46
Müller-Munk, Peter 51, 146
Museum of Modern Art (MoMA) 30, 47, 52, 134, 136, 165
mythology 67, 71

Nagy, Laszlo Moholy 53
Nairobi, University of 190
NASA 179
National Council for Diplomas in Design (NCDAD) 182
National Endowment for the Arts 178
The Needlewoman 113
Negus, Richard 182
Nelson, George 14, 15, 17, 43, 45, 49, 67–9, 74, 83, 105, 133, 134, 153, 154, 192, 193, 200
neocolonial development 179
New Yorker, The 106
New York state 49, 50, 180
New York Times 102, 106, 109, 167
New York World's Fair, 42–5, 54, 106, 138, 158
New Zealand 146
Nixon, Richard 178
Nixon, Sean 7, 11
Nordic 185
North Staffordshire 31
new, the 28
  New Deal 28, 42, 45, 47
  'new democracy' 31
  'new-ness' 43–4
  'new-wave' consultancies 65, 81, 85
  'New Woman' 97

'operation bootstrap' 43
obsolescence, 7, 44, 53, 56, 67, 104, 144, 155, 168, 179, 190
  see also planned obsolescence
occupation 10, 50, 51
Ogilvy, David 86, 103
Olin cellophane 73
oral history 16, 49, 115, 116

Packard, Vance 155, 184
Paepcke, Walter 81, 154, 169
Papanek, Victor 15, 178, 184–91, 193–4
Parkinson, C. Northcote 155, 156, 169
Parkinson's Law: the Pursuit of Progress (1958) 155
'party conversation', *SIA Journal*, (1952) 129, 131, 132
Peach, Harry 31
Pearson, David 145
periphery, the 187, 202
Petty France, London 116
Pick, Frank 34
Pilditch, James 85, 87, 160, 161, 162, 163, 166, 169, 173, 183
pioneers, 14, 43, 53, 66, 85, 86, 89, 100, 187, 193
  see also heroes, masculinity
pipe 38, 74
planned obsolescence 2, 28, 42, 43, 134, 153, 155, 157, 177, 178, 185, 192, 203
  see also obsolescence
Plaza Hotel, New York 167
polemic 16, 184, 190
Pond, Eddie 117, 183
postcolonial 190
pragmatism 143, 187
Procter, Jack 161
profession
  'creative profession' 2, 169
  'dangerous profession' 184, 178
  Declaration of Professional Behaviour 162, 163
  'inter-professional' and 'intra-professional' dynamics 148
  'new professions' 2–14, 27–9, 34, 43–5, 52, 55, 57, 67, 71, 73, 101, 104–7, 119, 120, 134, 135, 146, 149, 158, 204

Index

'older professions' 9, 29, 34, 142, 162
'semi-professional' status 2, 3
'traditional professions', 6, 192
professionalization
  crisis of professionalism 153, 158, 201, 202
  professional journals 128–129
  professional organization/society 4–8, 14,–16, 27–8, 33, 47–8, 52, 54, 128, 133, 165–73, 184, 190–3
  'pure professionalism' 164
  'The Professional Challenges of the 60s' 166
professionalism
  Professionalism and the Designer 160, 162
  professional conduct 1–18, 28, 127, 128, 141, 164
  professional expertise 1, 83, 89, 129, 157, 189
  professional ideal 1, 6, 10, 15, 28, 31, 34, 57, 135, 136, 156, 158, 162, 172, 177, 183, 192, 201, 203
  professional identity 1–18, 52–7, 65, 90, 98, 100–9, 132–8, 145–9, 165–9, 191
  professional self-image 3, 6, 27, 74, 89, 172
  professional status 2, 5, 6, 32–3, 51, 65–9, 74, 89, 105, 132–7, 145, 157, 162, 168, 180, 191
  The Professional Practice of Design 112, 114, 162, 192
Promotion 15, 85–6, 89, 105, 118, 161–4, 192
public, the
  public awareness 193
  public responsibilities 182
publicist, 16, 43, 56, 70, 76, 104, 107, 108, 113
  see also Reese, Betty
publicity, profession
  'undignified publicity' 141
public relations 3, 16, 42, 45, 78, 90, 104, 106–7, 132, 139, 155
Pulos, Arthur J., 9, 10, 11, 47, 71, 114, 134, 146, 148, 167, 168, 180, 181, 191, 194
Pursell, Carroll 97

'quasi-anthropological research paradigms' 189
questionnaire 72, 161

race 194, 202
Ray, Peter 163
Record of Designers 115, 140
Reid, John 160
Reilly, Paul 157, 160, 161
'The Relevance of Industrial Design' conference (1973) 189
Renton, Andrew 161
Renwick, William 165, 167
Reese, Betty 27, 107, 181
  see also publicist
rhetoric 2, 160, 185, 190, 201
Rockefeller, John D. 76, 169
Roosevelt, Franklin D. 42
Rose, Stuart 142, 162, 170
Royal Charter 178, 181, 182, 183
Royal College of Art (RCA) 78, 117, 127
Royal Designer for Industry (RDI) 33, 36, 75
Royal Institute of British Architects 158
Royal Society for the Promotion of Arts and Commerce (RSA) 30, 33, 64, 75
royalties *68*
Russell, Gordon 127, 165

Sakier, Georg 52, 67
salary 16, 67, 69, 143
San Francisco 48, 168, 184, 194
Sanu, 102
  see also Knoll Bassett, Florence
Scarfe, Laurence 129
Schreiber, Gabrielle (Gaby) 14, 76, 78–80, 96–103, 120
Second World War 5, 8, 27, 33, 34, 37, 44, 65, 83, 93, 154, 155
Shapira, Nathan 190, 193
Shaughnessy, Adrian 114
SIA Journal 129, 132, 140, 149, 158, 171
Silent Salesman, The 87, 160, 169
Simon, Herbert 31
Singer, Roger 166
Smirnoff vodka 80
social design 8, 198, 202

social good 177, 185
social media 202
'sociological wrapping' 65
social responsibility 8, 53, 76, 173, 178, 179, 180–5, 189, 192, 193, 194
'soft power' 145
Soviet Union 187
Sparke, Penny 11, 64, 85, 189
specialization
  see generalism vs specialism model
Spence, Basil 37
Spilman, Raymond 14, 16, 47, 49, 71, 81, 105, 143, 165, 166, 169, 173, 185, 200
staff designer see design disciplines
Stevens, Brooks 43, 69, 139, 144, 167, 184, 196
Stile Industria 112
Strel, Don 158
Stuart, William M 169
Studebaker 167, 196
student designers
  student behaviour report 127
  see also designers: 'young designers'
styling 12, 32, 51, 56
stylist 51, 69, 167, 181
Superstudio 158
Swing, Sally 107, 139, 144
Syracuse University 16, 167, 194

Tandy, John 142, 164
'taste-makers' 96, 149
Tatler 98, 101
tax 49, 50, 105
Taylor, Frederick Winslow 66
Teague vs. Graves (1941) 50
Teague, Walter Dorwin, 48–51, 67, 70, 71, 74, 88, 105, 136, 137, 141, 143, 165, 184, 187, 191
Thatcher, Margaret 182
Thomas, Zoë 6, 13
Thompson, Jane 95, 133
Thomson, E. M. 129
Times, The 80, 192, 103
Tomrley, Cycill 116–17, 121, 127, 132, 133, 149, 156
transatlantic relations 9
True, The Man's Magazine 144

Twemlow, Alice 15, 55, 62, 95, 109, 133, 155, 157

'unknowable woman' 13, 96
'user', the 173, 179, 188, 189
UNIDO 189, 190
United States President's Committee for Employment of the Handicapped 178
universalism, universality, 2, 10, 17

Van Doren, Harold 191, 48–50, 67, 76, 103, 114, 141
Victoria and Albert Museum (V&A), London 37, 38
Vienna 148, 185
Vietnam War 173
visibility 28, 33, 51, 57, 66, 75, 76, 80, 89, 97, 120, 141, 148, 173
Vogue 10, 43, 45, 80, 98, 100, 101, 138, 158

Wakeman, Alfred 144
Washington 179
The Waste Makers (1960) 155
Wedgwood, Josiah 77
Welsh, Vernon 157
Western designers 190
Western superiority 187
Wetmore, Robert 16, 113–16, 183
White House 179
whiteness 53
Whitney, Charles 68, 95, 133, 134
Whitney, John Hay 71, 169
Whitworth, Lesley 13, 38, 40, 96
Who's Who 97
Wilkins, Bridget 34
women
  women's work 95, 96, 106, 113
  'woman designer' 90–8, 100, 102, 103, 120
  'women's pages' 101
  see also femininity
Woolman-Chase, Edna, 45, 46
working from home, 14, 202
Wright, Russel, 53, 69, 71, 73, 102, 104, 139

Zachary, Stefan, 14, 183

Printed in the USA
CPSIA information can be obtained
at www.ICGtesting.com
JSHW070501031024
70945JS00003BA/4